AF720061

Next-generation Sequencing

Current Technologies and Applications

Edited by

Jianping Xu

Department of Biology
McMaster University
Hamilton, ON
Canada

Caister Academic Press

Copyright © 2014

Caister Academic Press
Norfolk, UK

www.caister.com

British Library Cataloguing-in-Publication Data
A catalogue record for this book is available from the British Library

ISBN: 978-1-908230-33-1 (hardback)
ISBN: 978-1-908230-95-9 (ebook)

Description or mention of instrumentation, software, or other products in this book does not imply endorsement by the author or publisher. The author and publisher do not assume responsibility for the validity of any products or procedures mentioned or described in this book or for the consequences of their use.

All rights reserved. No part of this publication may be reproduced, stored in a retrieval system, or transmitted, in any form or by any means, electronic, mechanical, photocopying, recording or otherwise, without the prior permission of the publisher. No claim to original U.S. Government works.

Cover design adapted from graphics courtesy of Jianping Xu

Printed and bound in Great Britain

Contents

Current books of interest

Biofuels: From Microbes to Molecules	2014
Human Pathogenic Fungi: Molecular Biology and Pathogenic Mechanisms	2014
Applied RNAi: From Fundamental Research to Therapeutic Applications	2014
Halophiles: Genetics and Genomes	2014
Phage Therapy: Current Research and Applications	2014
Bioinformatics and Data Analysis in Microbiology	2014
The Cell Biology of Cyanobacteria	2014
Pathogenic *Escherichia coli*: Molecular and Cellular Microbiology	2014
Campylobacter Ecology and Evolution	2014
Burkholderia: From Genomes to Function	2014
Myxobacteria: Genomics, Cellular and Molecular Biology	2014
Omics in Soil Science	2014
Applications of Molecular Microbiological Methods	2014
Mollicutes: Molecular Biology and Pathogenesis	2014
Genome Analysis: Current Procedures and Applications	2014
Bacterial Toxins: Genetics, Cellular Biology and Practical Applications	2013
Bacterial Membranes: Structural and Molecular Biology	2014
Cold-Adapted Microorganisms	2013
Fusarium: Genomics, Molecular and Cellular Biology	2013
Prions: Current Progress in Advanced Research	2013
RNA Editing: Current Research and Future Trends	2013
Real-Time PCR: Advanced Technologies and Applications	2013
Microbial Efflux Pumps: Current Research	2013
Cytomegaloviruses: From Molecular Pathogenesis to Intervention	2013
Oral Microbial Ecology: Current Research and New Perspectives	2013
Bionanotechnology: Biological Self-assembly and its Applications	2013
Real-Time PCR in Food Science: Current Technology and Applications	2013
Bacterial Gene Regulation and Transcriptional Networks	2013
Bioremediation of Mercury: Current Research and Industrial Applications	2013
Neurospora: Genomics and Molecular Biology	2013
Rhabdoviruses	2012
Horizontal Gene Transfer in Microorganisms	2012
Microbial Ecological Theory: Current Perspectives	2012
Two-Component Systems in Bacteria	2012

Full details at www.caister.com

Contributors

Tudor Constantin
Eureka Genomics
Hercules, CA
USA

tudor@eurekagenomics.com

Christoph Dieterich
Bioinformatics in Quantitative Biology
Berlin Institute for Medical Systems Biology at the Max Delbrück Centre for Molecular Medicine
Berlin
Germany

christoph.dieterich@mdc-berlin.de

Viacheslav Y. Fofanov
Eureka Genomics
Sugar Land, TX
USA

vyfofanov@eurekagenomics.com

Leighton Folkes
School of Computing Sciences
University of East Anglia
Norwich
UK

l.folkes@uea.ac.uk

Jack A. Gilbert
Department of Ecology and Evolution
University of Chicago
Chicago, IL
USA

gilbertjack@gmail.com

Johannes B. Goll
J. Craig Venter Institute
Rockville, MD
USA

johannes.goll@gmail.com

Tao Jiang
BGI-Shenzhen
Beishan Industrial Zone
Shenzhen
China

jiangtao@genomics.cn

Yuling Jiao
State Key Laboratory of Plant Genomics
Institute of Genetics and Developmental Biology
Chinese Academy of Sciences
Beijing
China

yljiao@genetics.ac.cn

Heather Koshinsky
Eureka Genomics
Hercules, CA
USA

heather@eurekagenomics.com

Konstantinos Krampis
J. Craig Venter Institute
Rockville, MD
USA

kkrampis@jcvi.org

Peter Larsen
Argonne National Laboratory
Argonne, IL
USA

plarsen@anl.gov

Jun Li
BGI-Shenzhen
Beishan Industrial Zone
Shenzhen
China

lijun1@genomics.cn

Ramana Madupu
J. Craig Venter Institute
Rockville, MD
USA

rmadupu@jcvi.org

Daniel Mapleson
The Genome Analysis Centre
Norwich Research Park
Norwich
UK

daniel.mapleson@tgac.ac.uk

Barbara A. Methé
J. Craig Venter Institute
Rockville, MD
USA

bmethe@jcvi.org

Irina Mohorianu
School of Computing Sciences
University of East Anglia
Norwich
UK

i.mohorianu@uea.ac.uk

Vincent Moulton
School of Computing Sciences
University of East Anglia
Norwich
UK

v.moulton@uea.ac.uk

Karen E. Nelson
J. Craig Venter Institute
Rockville, MD
USA

kenelson@jcvi.org

Zemin Ning
The Wellcome Trust Sanger Institute
Cambridge
UK

zn1@sanger.ac.uk

Fatih Ozsolak
Helicos BioSciences Corporation
Cambridge, MA
USA

fatihozsolak@gmail.com

Helio Pais
The Weatherall Institute of Molecular Medicine
University of Oxford John Radcliffe Hospital
Oxford
UK

helio.pais@imm.ox.ac.uk

Michael Piechotta
Bioinformatics in Quantitative Biology
Berlin Institute for Medical Systems Biology at the Max Delbrück Centre for Molecular Medicine
Berlin
Germany

michael.piechotta@mdc-berlin.de

Chandra Shekhar Pareek
Functional Genomics in Biological and Biomedical Research
Centre for Modern Interdisciplinary Technologies and Faculty of Biology and Environment Protection
Nicolaus Copernicus University
Torun
Poland

chandra.s.pareek@gmail.com

Matthew Stocks
School of Computing Sciences
University of East Anglia
Norwich
UK

matthew.stocks@uea.ac.uk

Sebastian Szpakowski
J. Craig Venter Institute
Rockville, MD
USA

shpakoo@gmail.com

Jun Wang
BGI-Shenzhen
Beishan Industrial Zone
Shenzhen
China

wangj@genomics.cn

Ying Wang
State Key Laboratory of Plant Genomics
Institute of Genetics and Developmental Biology
Chinese Academy of Sciences
Beijing
China

y.wang@genetics.ac.cn

Andreas Wilke
Argonne National Laboratory
Argonne, IL
USA

wilke@mcs.anl.gov

Xu Yang
BGI-Shenzhen
Beishan Industrial Zone
Shenzhen
China

yangxu@genomics.cn

Kai Ye
The Genome Institute
Washington University
St. Louis, MO
USA

liangkaiye@gmail.com

Preface

A defining feature of the human race is our curiosity and drive to explore and understand the unknown. Examples of our searches for unknowns are found throughout our recorded history. Understanding ourselves and the organisms around us as well as those in the past (and sometimes in the future) has been an eternal topic of these quests. Centuries of research have identified that biological phenomena are governed by the inherited features of organisms (their genomes and epigenomes), their biotic and abiotic environments, and the spatial–temporal interactions between their (epi-)genomes and environmental factors. Advances in technology have played critical roles in increasing our understanding. This is especially true with the development and application of next-generation DNA sequencing (NGS) technologies. Over the past decade, NGS technologies have brought unprecedented opportunities and insights into understanding all types of biological phenomena, from the causes of cancer to the diagnosis of viral pathogens, from the diversities of life on the sea floor to the adaptations of life at high altitudes, from assembling the genomes of unknown organisms to constructing the genome map of extinct animals, from helping debunk the nature vs. nurture myth debate about human intelligence to identifying the causes and consequences of the diversity of life history strategies, and from the genomic changes of domesticated plants and animals to the impacts of microbial flora on the health of mushrooms, plants, animals, and humans.

DNA sequencing is the process of determining the precise order of nucleotides within a DNA molecule. It includes any method or technology that is used to determine the order of the four bases—adenine (A), guanine (G), cytosine (C), and thymine (T)—in a strand of DNA. Though the structure of DNA was established as a double helix in 1953, it was not until the 1970s that fragments of DNA could be reliably analyzed for their sequence in the laboratory. The first generation of DNA sequencing techniques included the chemical sequencing method developed by Allan Maxam and Walter Gilbert in 1973 and the chain termination method by Frederick Sanger in 1977. The chain termination method (commonly called the Sanger method) was the basis for the first semi-automated DNA sequencing machine in 1986, which one-year later was developed into a fully automated sequencing machine, the ABI 370, by Applied Biosystems. Several improved models were subsequently developed and used to sequence the first bacterial genome in 1995, the first eukaryote genome in 1996, and the first draft sequence of the human genome in 2001.

Since the mid-1990s, new DNA sequencing methods and instruments have been developed. The first of these new techniques was the pyrosequencing method by Pål Nyrén and Mostafa Ronaghi in 1996; quickly followed by the DNA colony sequencing by Pascal Mayer and Laurent Farinelli in 1997; and the massively parallel signature sequencing (MPSS) by Lynx Therapeutics in 2000. Though MPSS machines were produced and marketed, it was not sold to independent laboratories. The first commercial 'next-generation' sequencing machine was by 454 Life Sciences in 2004 based on the pyrosequencing principle.

This machine reduced sequencing cost by 6 folds over the most advanced sequencers using the Sanger method. Other sequencing methods and commercial platforms soon followed, including Illumina (Solexa), SOLiD, Ion Torrent, DNA Nanoball, Heliscope, and Single molecule real time (SMRT) sequencing. In addition, several methods are in development, including Nanopore DNA sequencing, Sequencing by hybridization, Sequencing with mass spectrometry, Microfluidic Sanger sequencing, Microscopy-based techniques, RNAP sequencing, and *in vitro* virus high-throughput sequencing.

These new developments reflect in part our increased understanding of DNA chemistry, innovations in highly sensitive and high throughput detection methods, and the development of computational tools for acquiring, storing and analyzing the massive amount of DNA sequence data. Additional impetus for more accurate and more efficient DNA sequencing methods are driven by the applications and promises of what NGS technologies can bring to improve the health of not only humans, but also agricultural crops, trees, animals, the environment, and indeed, the entire global biosphere. Furthermore, in October 2006, the X Prize Foundation established the Archon X Prize to promote the development of new genome sequencing technologies. Specifically, it will award $10 million to 'the first Team that can build a device and use it to sequence 100 human genomes within 10 days or less, with an accuracy of no more than one error in every 100,000 bases sequenced, with sequences accurately covering at least 98% of the genome, and at a recurring cost of no more than $10,000 (US) per genome.' Since this PRIZE was first launched in 2006, DNA sequencing technologies both in terms of speed and reduction in costs continue to advance. However, no current human genome sequence is fully complete, fully accurate, or capable of inferring all rearrangements and chromosome phasing (haplotype). In addition, highly repetitive genomic regions remain difficult to sequence. The 100 human genomes to be sequenced in this competition are donated by 100 centenarians (ages 100 or older) from all over the world, known as the '100 Over 100'. The reasons for choosing this group of individuals were to help identify the 'rare alleles' that might protect against disease and promote longevity. Though the prize competition was recently abandoned owing to its stringent criteria and insufficient entries, developing such a system remains the ultimate objective for the sequencing community.

The objective of this book is to provide a broad overview of NGS technologies and the applications of NGS in selected fields of biomedical research. The goal is not to provide an exhaustive review but rather to highlight the broad and representative areas of what NGS technologies can revolutionize our understanding of a diversity of issues. The book is composed of 10 chapters, each by one or a group of prominent experts. Chapter 1 provides an overview of NGS platforms by Professor Chandra Pareek. In Chapter 2, Dr Fatih Ozsolak describes a specific type of sequencing technology that performs single-molecule direct DNA, cDNA and RNA sequencing at attomole-level. Chapters 3 to 8 deal with the processing and analyses of NGS data to address specific types of issues, including SNP assessment based on draft genomes by Drs Piechotta and Dieterich (Chapter 3); large-scale small RNA identification and analyses by Drs Mapleson *et al.* (Chapter 4); rare variant detection by Drs Fofanov, Constantin and Koshinsky (Chapter 5); identifying breakpoints for insertions and deletions by Drs Ye and Ning (Chapter 6); inferring novel insights from human exome re-sequencing data by Drs Li, Jiang, Yang and Wang (Chapter 7); and identifying genes and genetic pathways govern plant development by Drs Wang and Jiao (Chapter 8). These six chapters cover a diversity of topics and deal with representatives of major forms of identified organisms such as microbes, plants, and animals including humans. In contrast, Chapters 9 and 10 review and discuss how NGS technologies have influenced and will continue to impact our understandings of mostly unknown microbes in natural environments. Metagenomics directly investigate environmental DNA samples that contain genetic materials from both known organisms (a relatively small proportion) but predominantly organisms of unknown identities. Chapter 9 by Drs Wilke, Larsen and Gilbert reviewed recent advances and presented a framework on how NGS technologies have impacted and will continue to drive research on microbial communities at both structural and functional levels. Chapter 10 by Nelson *et al.* summarized the main findings of metagenomics research results, with emphases on the need for a systems biology perspective and the advances of our understanding of the human microbiome, one of the most intensely investigated and best understood environmental genomes. All chapters in the book discussed the needs for standardized NGS data storage, analyses, and information-sharing platforms.

I feel extremely honored to be able to work with and learn from this group of fantastic scientists. I hope you will enjoy reading the chapters as much as I have been. This book would not have been possible without the hard work of all contributors and I thank each of them for their contribution and for making this book a reality. I also want to thank Hugh Griffin at Horizon Scientific Press, Caister Academic Press, for his unwavering support and patience throughout the process.

Jianping Xu

An Overview of Next-generation Genome Sequencing Platforms

1

Chandra Shekhar Pareek

Abstract

The next-generation genome sequencing platforms are principally based on the immobilization of the DNA samples onto a solid support, cyclic sequencing reactions using automated fluidics devices, and detection of molecular events by imaging. This chapter reviews the platform-specific sequencing chemistries, instrument specifications, and general workflows and procedures of both the popular second generation sequencing platforms: GS FLX by 454 Life Technologies/ Roche Applied Science, genome analyser by Solexa/Illumina, SOLiD by Applied Biosystems, and the third generation sequencing platforms: semiconductor sequencing by Ion Torrents/Life Technologies, PacBio RS by Pacific Biosciences, HeliScope™ single molecule sequencer by Helicos Bioscience Cooperation, Nanopore sequencing by Oxford Nanopore Technologies.

Introduction

The first successful DNA sequencing methods were introduced in the 1970s using the terminally labelled DNA molecules to cleave chemically at specific base fragments [Maxam and Gilbert (1977) method], or utilizing the dideoxynucleotide analogues as specific chain-terminating inhibitors of DNA polymerase [Sanger *et al.* (1977) method]. Since then, the Sanger dideoxy sequencing method has been the method of choice for DNA sequencing and recognized as the first-generation sequencing (FGS) technology. In 2005, an era of next-generation genome sequencing (NGS) began with the development of second generation sequencing (SGS) platforms (Margulies *et al.*, 2005), and third generation sequencing (TGS) platforms (Eid *et al.*, 2009). During the years 2005–2012, development of SGS and TGS platforms has revolutionized the genome sequencing technologies in unprecedented manner and a large number of peer-reviewed papers have been published based on data generated from NGS (Mardis, 2008, 2011, 2012; Rusk, 2009; Metzker, 2010; Ozsolak and Milos, 2011a; Schadt *et al.*, 2010; Pareek *et al.*, 2011). In general, the NGS platforms share a common technological feature i.e. massively parallel sequencing of clonally amplified or single DNA molecules that are spatially separated in a SGS platforms (also termed as non-Sanger ultra-high-throughput sequencing platforms) and TGS platforms (also termed as flow cell platforms of single-molecule sequencing and semiconductor sequencing). This book chapter provides a comprehensive description of three SGS platforms and four TGS platforms, particularly towards their platform-specific sequencing chemistries, instrument specifications, and general workflows and procedures.

Second generation sequencing platforms

The SGS platforms can be defined as a combination of a synchronized reagent wash of nucleoside triphosphates (NTPs) with a synchronized optical detection method. A major advantage of the SGS platforms over the FGS platform is replacing the *in vivo* cloning or ex vivo PCR amplification prior to sequencing reaction with an in situ PCR-based amplification. In general, SGS platforms rely on two principles: polymerase-based clonal replication of single DNA molecules spatially separated on a solid support matrix (bead or planar surface) and cyclic sequencing chemistries such as sequencing by ligation or sequencing by synthesis, including pyrosequencing and reversible chain termination. Therefore, the basic rule behind SGS platforms is to randomly fragment DNA or RNA into smaller pieces and construct a DNA or cDNA library. By making them smaller, micro-RNAs (miRNAs) can be used directly to make cDNA libraries for direct sequencing. Finally, the constructed libraries are sequenced at a high coverage and the sequenced reads are then assembled and/or mapped onto the reference genome of the species. Thus, these platforms allow highly streamlined sample preparation steps prior to DNA sequencing, which provides a significant time saving and a minimal requirement for associated equipment in comparison to the highly automated, multistep pipelines necessary for clone-based high-throughput FGS platforms. Moreover, each SGS platform seeks a unique combination of specific protocols to amplify single strands of a DNA fragment library and perform sequencing reactions on the amplified strands. The specific combinations distinguish one technology from another and determine the type of data produced from each platform. Three commercially available SGS platforms widely used in global genomics researches are the Roche Applied Science/454 (the Junior and GS-FLX$^+$ system), Illumina/Solexa (genome analyser system), and Applied Biosystems/Life Technologies (SOLiD system). Each platform embodies a complex interplay of enzymology, chemistry, high-resolution optics, hardware, and software engineering. Below I describe each of the three platforms.

Roche SGS platforms

Available instruments

The Roche Applied Science (www.roche-applied-science.com/) NGS system was first commercialized in late 2004, under 454 Life Sciences (http://www.my454.com/) based on the pyrosequencing technology. The platform provides a bench-top personal genome sequencer GS Junior Titanium directly to the laboratory and two high-throughput long read sequencing systems, i.e. 454 FLX$^+$ and 454 FLX Titanium, that are well suited for larger genomic projects. The highlights and important features of these instruments are illustrated in Table 1.1.

Sequencing chemistry

Both the Roche Junior and GS-FLX$^+$ SGS platform sequencing chemistry is based on sequencing-by-synthesis using pyrosequencing, where the release of pyrophosphate upon nucleotide incorporation results in luminescent signal output (Ronaghi *et al.*, 1998, 1999). In this approach, DNA synthesis is performed within a complex reaction that includes enzymes (ATP sulphurylase and luciferase) and substrates (adenosine 5′-phosphosulphate and luciferin) such that

Table 1.1 Specification of Roche SGS instrument

Instrument	Run time (hours)	Read length (bp)	Reads/run (millions)	Yield (mb/run)	Reagent cost/run[a] ($)	Reagents cost/mb[b] ($)	Minimum unit cost (%r un)[b] ($)	Error type	Error rate (%)	Purchase cost ($)
454 GS Jr. Titanium	10	400	0.10	50	1100	22	1500 (100%)	indel	1	108 000
454 FLX Titanium	10	400	1	500	6200	12.4	2000 (10%)	Indel	1	500 000
454 FLX$^+$	18–20	800–1000	1	900	6200	7	2000 (10%)	Indel	1	300 000

[a]Includes all stages of sample preparation for a single sample (i.e. library preparation through sequencing).
[b]Typical full cost (i.e. including labour, service contract, etc.) of the smallest generally available unit of purchase at an academic core laboratory provider for the longest available read (and percentage of reads relative to a full run, rounded to the nearest whole percentage).

the pyrophosphate group (PPi) released upon incorporation of the nucleotides by chemiluminescence, resulting in the production of detectable light (Fig. 1.1). Therefore, when a new nucleotide is incorporated in a growing DNA chain through the activity of DNA polymerase, pyrophosphate is generated in a stoichiometric manner resulting in the production of ATP. The ATP produced drives the enzymatic conversion of luciferin with the associated emission of photons. This is followed by an apyrase wash to remove unincorporated nucleotide. Subsequent to the clean up of the reaction compounds, a new cycle of reactants are introduced and thus the incorporation of specific nucleotides is measured in a sequential manner.

General workflow

In Roche SGS platforms, the sample preparation process is much simplified compared with Sanger sequencing. There are three sample preparation steps in the Roche SGS platforms: fragmentation of genomic DNA by nebulization, clonal amplification of the target DNA on micron-sized beads using an emulsion-based PCR method (emPCR), and pyrosequencing.

Preparation of the DNA library

This step involves random fractionation of genomic DNA (gDNA) into smaller single-stranded template DNA (sstDNA) fragments (300–1000 bp) that are subsequently polished (blunted) and then, short adaptors (A and B) are nick-ligated onto both ends. These adaptors provide priming sequences for the amplification and sequencing of the sample-library fragments as well as act as special tags. The adaptor B contains a 5′-biotin tag that enables immobilization of the library onto streptavidin-coated beads. After nick repair, the non-biotinylated strand is released and used as an sstDNA library. The sstDNA library is assessed for its quality and the optimal DNA copy number needed per bead for emPCR™ is determined by titration (Fig. 1.2).

The emPCR™

The emPCR™ is carried out for sstDNA library amplification, with water droplets containing one bead and PCR reagents immersed in oil. The single stranded DNA library is immobilized by hybridization onto primer-coated capture beads. The process is optimized to produce beads where a single library fragment is bound to each bead. The bead-bound library is emulsified along with the amplification reagents in a water-in-oil mixture. In Roche SGS platform, each bead is captured within its own microreactor where the independent clonal amplification takes place. After amplification, the microreactors are broken, releasing the DNA-positive beads for clonal amplification of DNA fragments (Fig. 1.3). This amplification is necessary to obtain sufficient light signal intensity for reliable detection in the next step of sequencing-by-synthesis reaction. Finally, when PCR amplification cycles are completed and after denaturation, each bead with its one amplified fragment is placed at the top end of an etched fibre in an optical fibre chip, created from glass fibre bundles.

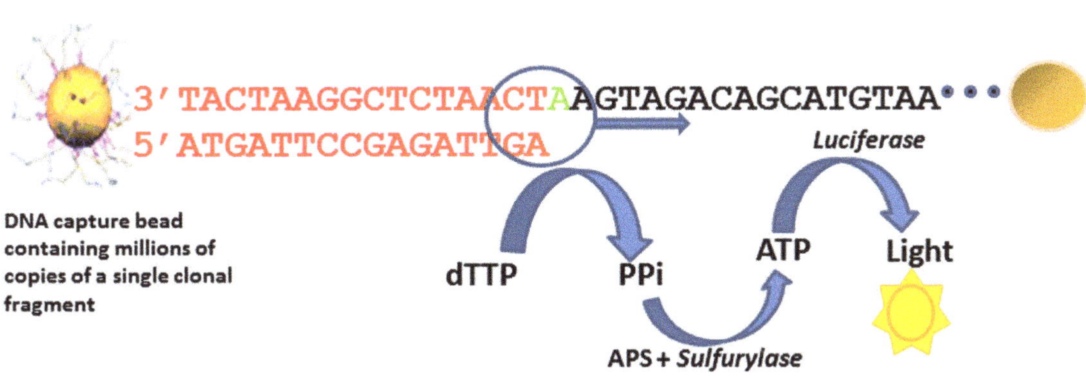

Figure 1.1 Pyrosequencing chemistry of Roche SGS platform.

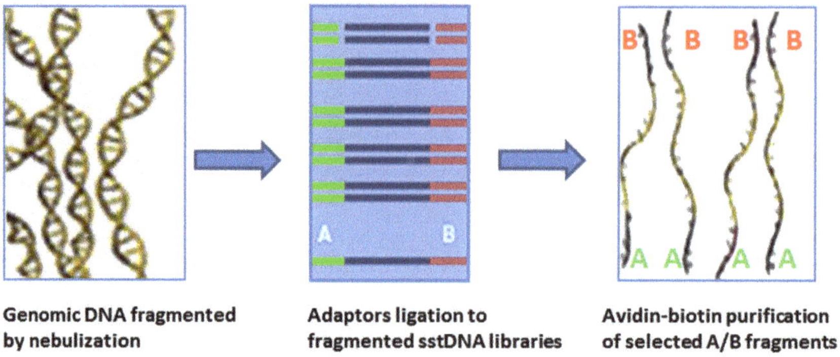

Figure 1.2 Preparation of the sequencing DNA library in Roche SGS platform.

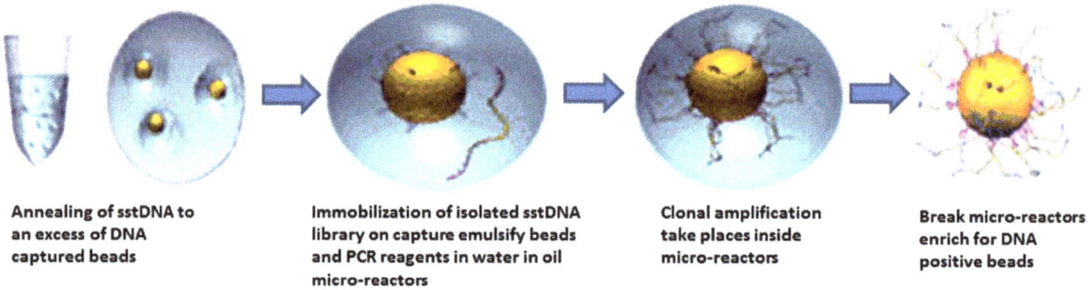

Figure 1.3 The emulsion PCR in Roche SGS platform.

Pyrosequencing

The Roche SGS platform uses a picotitre plate (Margulies *et al.*, 2005), a solid phase support containing over a million picoliter volume wells. The dimensions of the wells are such that only one bead is able to enter each well on the plate. Sequencing chemistry flows through the plate and insular sequencing reactions take place inside the wells. The picotiter plate can be compartmentalized up to 16 separate reaction entities using different gaskets. In the pyrosequencing step, the DNA beads are layered onto a PicoTiterPlate (PTP) device and then the beads are deposited into the wells, followed by enzyme beads and packing beads. The enzyme beads contain sulphurylase and luciferase, which are key components of the sequencing reaction, while the packing beads ensure that the DNA beads remain positioned in the wells during that sequencing reaction. The fluidics subsystem delivers sequencing reagents that contain buffers and nucleotides by flowing them across the wells of the plate (Fig. 1.4). Nucleotides are flowed sequentially in a specific order over the PTP device. When a nucleotide is complementary to the next base of the template strand, it is incorporated into the growing DNA strand by the polymerase. The incorporation of a nucleotide releases a pyrophosphate moiety. The sulphurylase enzyme converts the pyrophosphate molecule into ATP using adenosine phosphosulphate. The ATP is hydrolysed by the *luciferase* enzyme using luciferin to produce oxyluciferin and gives off light. The light emission is detected by a CCD camera, which is coupled to the PTP device. The intensity of light from a particular well indicates the incorporation of nucleotides. Across multiple cycles, the pattern of detected incorporation events reveals the nucleotide sequence of individual templates represented by individual beads. The sequencing is asynchronous in that some features may get ahead or behind other features depending on their sequence relative to the order of base addition. Raw reads processed by the 454 platforms are screened by various quality

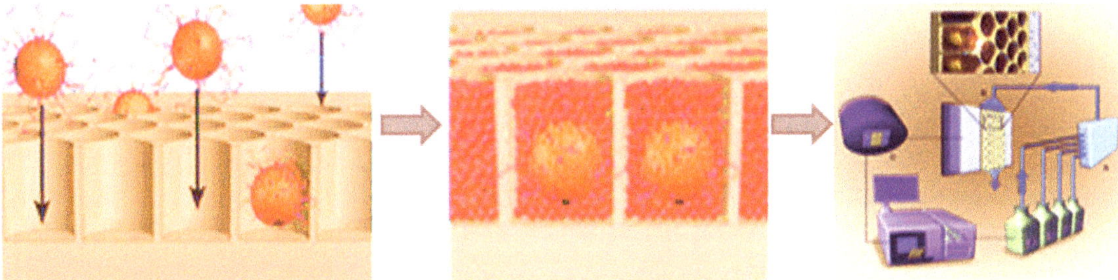

Loading of amplified sstDNA libraries beads to 44μm diameter well of PTP plate

PPiase **+ enzyme + packing beads loading of a single cloned amplified sstDNA bead is deposited per well**

Pyro-sequencing

Figure 1.4 Pyrosequencing in Roche SGS platforms.

filters to remove poor-quality sequences, mixed sequences (more than one initial DNA fragment per bead), and random sequences generated without an initiating template sequence.

Illumina (Solexa) SGS platforms

Available instruments

The Illumina (Solexa) SGS platform was commercialized in 2006, with Illumina acquiring Solexa in early 2007 (www.illumina.com). This platform provides the following instruments: GAIIx, HiSeq 2000 (2010), HiSeq 2000 v.3 (2011) and bench-top personal genome sequencer MiSeq 2000 (2011). The highlights and important features of these instruments are presented in Table 1.2.

Sequencing chemistry

The Illumina SGS platform utilizes sequencing-by-synthesis chemistry, with novel reversible terminator nucleotides for the four bases each labelled with a different fluorescent dye, and a

Table 1.2 Specifications of Illumina SGS instruments

Instrument	Run time (days)	Read length (bp)	Reads/ run (millions)	Yield (Gb/ run)	Reagent cost/ run[a] ($)	Reagents cost/mb ($)	Minimum unit cost (% run) [b] ($)	Error type	Error rate (%)	Purchase cost ($)
MiSeq	1	150+ 150	3.4	2	750	0.74	~1000 (100%)	substitutions	>0.1	125 000
GAIIx	14	150+ 150	320	96	11,524	0.12	3200 (14%)	substitutions	>0.1	256 000
Hiseq 1000	8	100+ 100	500	100× 2	10,220[c]	0.10	3000 (12%)	substitutions	>0.1	425 000
HiSeq 2000	8	100+ 100	1000	200× 2	20,120[c]	0.10	3000 (6%)	substitutions	>0.1	654 000
HiSeq 2000 v3	10	100+ 100	<3000	<600	23,470[c]	>0.04	~3500 (6%)	substitutions	>0.1	700 000
Illumina iScanSQ	8	100+ 100	250	50	10,220	020	3000 (14%)	substitutions	>0.1	700 000

[a]Includes all stages of sample preparation for a single sample (i.e. library preparation through sequencing).
[b]Typical full cost (i.e. including labour, service contract, etc.) of the smallest generally available unit of purchase at an academic core laboratory provider for the longest available read (and percentage of reads relative to a full run, rounded to the nearest whole percentage).
[c]More reads are obtained than is needed from any single sample within most experiments, but the value illustrates the costs.

special DNA polymerase enzyme able to incorporate them. Illumina sequencing works on the principle of reversible termination with each sequencing cycle involving the addition of DNA polymerase and a mixture of four differently coloured reversible dye terminators followed by imaging of the flow cell. The terminators are then unblocked and the reporter dyes cleaved and washed away. Following sequencing from a single end of the template, paired-end sequencing can be achieved by sequencing from an alternate primer on the reverse stand of the template molecule (Fig. 1.5).

General workflow

The general workflow of Illumina SGS platform is based on immobilizing linear sequencing library fragments using solid support amplification followed by DNA sequencing using the fluorescent reversible terminator nucleotides.

Preparation of the DNA library

The DNA/RNA template is fragmented by a hydrodynamic process, then the fragments are blunt ended and phosphorylated, followed by ligation of DNA fragments at both ends to adapters using the single base 'A' overhang nucleotide to the 3′ ends and single-base 'T' overhang nucleotide to the 3′ ends. After denaturation, DNA fragments are immobilized at one end on a solid support-flow cell. The surface of the support is coated densely with the adapters and the complementary adapters (Fig. 1.6). Each single-stranded fragment, immobilized at one end on the surface, creates a 'bridge' structure by hybridizing with its free end to the complementary adapter on the surface of the support.

Cluster generation and solid support bridge PCR

The cluster generation on a flow cell is performed using the Illumina cBot cluster generation system. The PCR amplification takes place in solid support flow cell containing a mixture of PCR amplification reagents, the densely coated adapters and the complementary adapters. PCR is performed by the repeated thermal cycles of denaturation, annealing, and extension to exponentially amplify DNA molecules. The amplification is accomplished on a solid support using immobilized primers and in isothermal conditions using reagent flush cycles of denaturation, annealing, extension, and wash. During amplification, each single stranded fragment that is immobilized at one end on the surface creates a bridge structure by hybridizing with its free end to the complementary adapter on the surface of the flow cell (Fig. 1.7). The adapters on the surface also act as primers for subsequent PCR amplifications. After several PCR cycles, about 1000 copies of single-stranded DNA fragments are created on the surface, forming a surface-bound DNA clusters on each flow cell lane. DNA clusters are finalized for sequencing by unbinding the complementary DNA strand to retain a single molecular species in each cluster, in a reaction called linearization, followed by blocking the free 3′ ends of the clusters and hybridizing a sequencing primer (Fig. 1.7).

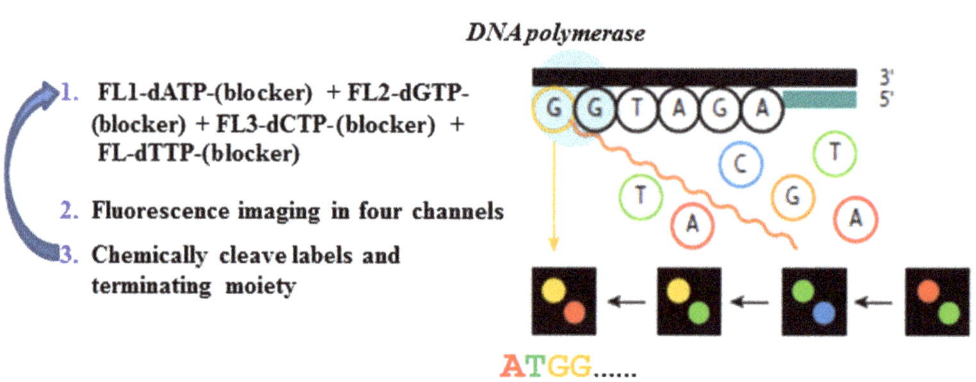

Figure 1.5 Illumina SGS platform sequencing-by-synthesis chemistry.

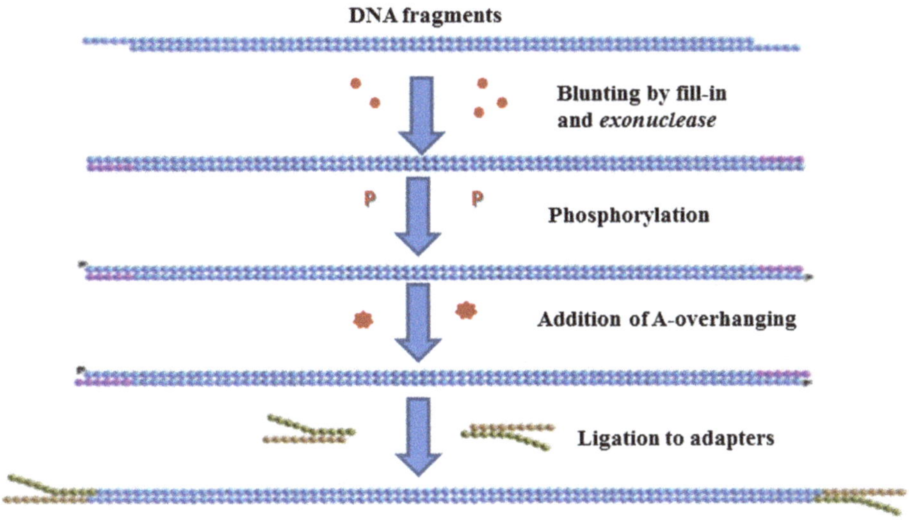

Figure 1.6 Preparation of the DNA library in Illumina SGS platform.

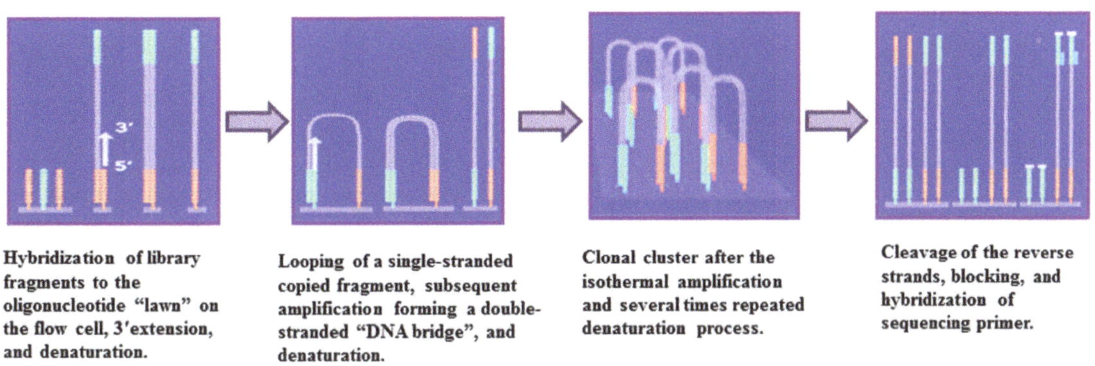

Figure 1.7 Illustration of cluster generation on a flow cell.

Sequencing-by-synthesis using fluorescently labelled reversible terminator nucleotides

The reaction mixture for the sequencing-by-synthesis is supplied onto the surface, which contains four reversible terminator nucleotides, each labelled with a different fluorescent dye. After incorporation into the DNA strand, the terminator nucleotide as well as its position on the support surface are detected and identified via its fluorescent dye by the CCD camera. The terminator group at the 3′-end of the base and the fluorescent dye are then removed from the base and the synthesis cycle is repeated. This series of steps continues for a specific number of cycles, as determined by user-defined instrument settings. A base-calling algorithm assigns sequences and associated quality values to each read and a quality-checking pipeline evaluates the Illumina data from each run, removing poor-quality sequences (Fig. 1.8).

The life technologies/applied biosystems (SOLiD) SGS platforms

Available instruments

The Life Science/Applied Biosystems (http://www.appliedbiosystems.com/) SGS platform was first commercialized in 2006 (McKernan *et al.*, 2009), acquiring the patent of sequencing by ligation technology (Macevicz, 1998). Their platforms include three high-throughputs long

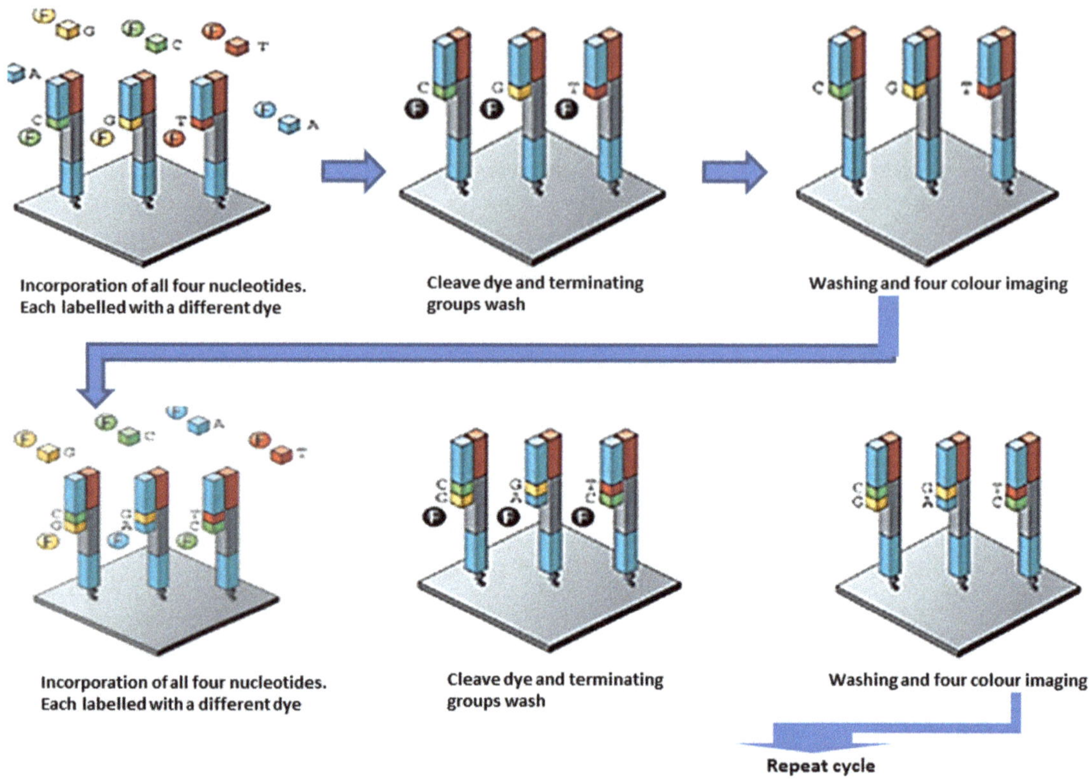

Figure 1.8 The four-colour cyclic reversible termination (CRT) sequencing by-synthesis method using solid-phase-amplified template clusters.

read sequencing systems, i.e. SOLiD-4, SOLiD-5500 PI and SOLiD-5500 XL. The highlights and important features of these instruments are presented in Table 1.3.

Sequencing chemistry

The SOLiD SGS platforms' sequencing chemistry is unique and is based on the principle of sequencing by oligo ligation and detection technology, using fluorescent probes from a common universal primer (Valouev *et al.*, 2008).

Key featuring points of SOLiD SGS chemistry (Fig. 1.9) are:

1. The annealing of a universal primer is performed with the addition a library of 1,2-probes.
2. The unique ligation of a probe to the primer can be performed bi-directionally from either its 5′-PO_4 (labelled probes) or 3′-OH (degenerated) end which allows the appropriate conditions to perform the selective hybridization and ligation of probes to complementary positions.
3. After the four-colour imaging, the ligated 1,2-probes are chemically cleaved with silver ions to generate a 5′-PO_4 group.
4. The probes also contain inosine bases (z) to reduce the complexity of the 1,2-probe library and a phosphorothiolate linkage between the fifth and six nucleotides of the probe sequence, which is cleaved with silver ions. Other cleavable probe designs include RNA nucleotides and internucleosidic phosphoramidates, which are cleaved by *ribonucleases* and acid, respectively.
5. The SOLiD sequencing by ligation cycle is repeated ten times, including an end cycle of stripping the extended primer and in total, four such ligation rounds are performed.

Table 1.3 Specifications of SOLiD SGS instruments

Instrument	Run time (days)	Read length (bp)	Reads/ run (millions)	Yield (Gb/ run)	Reagent cost/run[a] ($)	Reagents cost/m[b] ($)	Minimum unit cost (% run) [b] ($)	Error type	Error rate (%)	Purchase cost ($)
SOLiD-4	12	50× 35 PE	>840[d]	71	6200	7	2000 (10%)	A-T Bias, Substitutions	>0.06	475 000
SOLiD-5500 PI*	8	75× 35 PE	>700[d]	77	6101	0.08	2000 (12%)	A-T Bias, Substitutions	>0.06	495 000
SOLiD-5500 xl*	8	75× 35 PE	>1410[d]	155	10 503[c]	0.07	2000 (12%)	A-T Bias, Substitutions	>0.06	595 000

[a]Includes all stages of sample preparation for a single sample (i.e. library preparation through sequencing).
[b]Typical full cost (i.e. including labour, service contract, etc.) of the smallest generally available unit of purchase at an academic core laboratory provider for the longest available read (and percentage of reads relative to a full run, rounded to the nearest whole percentage).
[c]More reads are obtained than is needed from any single sample within most experiments, but the value illustrates the costs.
[d]Mappable reads [number of raw high-quality reads (as reported for all other platforms) is higher].
*Information based on company sources alone (independent data not yet available). PE, paired end.

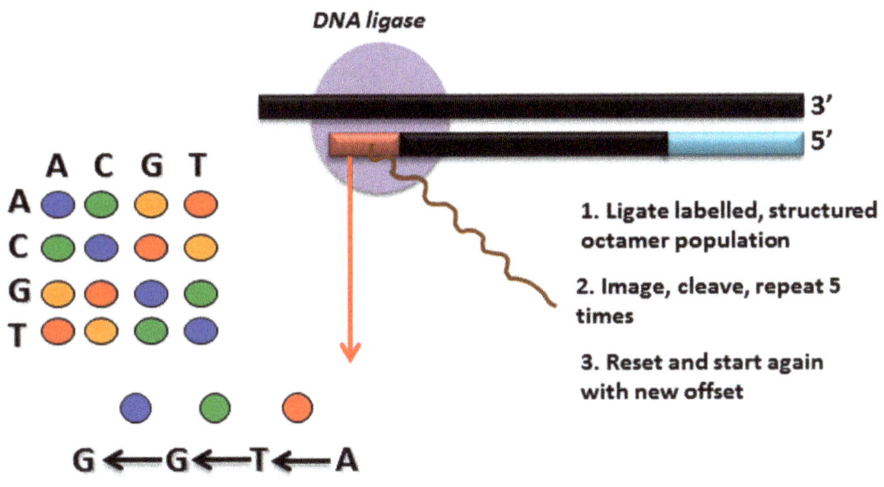

Figure 1.9 A four-colour sequencing by the ligation method using the oligonucleotide ligation detection (SOLiD) SGS platform.

6 The 1,2-probes are designed to interrogate the first (x) and second (y) positions adjacent to the hybridized primer in such way that the 16 dinucleotides are encoded by four colour dyes.

General workflow

Preparation of the DNA library

Similar to Roche SGS platforms, the sample preparation step for SOLiD platforms includes the generation of standard shotgun fragment library, followed by end-repair and phosphorylation and ligation of P1 and P2 adapters. Since *DNA ligases* are uniquely discriminating at both 5′ and 3′ sides of a ligation junction, the paired-end reads libraries can be generated from standard shotgun fragments simply by performing the sequencing reaction from the P2 adapter after one read. However, for the longer insert paired-end libraries (600 bp to 6 kb), a circularization protocol is utilized, in which, the adapters are ligated to the ends of the sheared, end-repaired DNA, and following size selection, fragments are circularized by ligation to an additional biotinylated adapter. Thus, ligation is prevented at one side of the adapter in each strand, leaving a nick in each strand. A mixture of nucleotides and DNA polymerase I is then used to translate these nicks into the template region at either side of the adapter, and the fragments containing the adapter are excised by *nuclease* digestion and are captured with 1 μm paramagnetic beads (Dressman *et al.*, 2003). In sample preparations, the libraries can be constructed by any method that gives rise to a mixture of short, adaptor-flanked fragments, though much effort with this system has been put into protocols for mate-paired tag libraries with controllable and highly flexible distance distributions (Shendure *et al.*, 2005).

Emulsion PCR and substrate preparation

DNA fragments for SOLiD SGS platforms are first amplified on the surfaces of 1 μm magnetic beads to provide sufficient signal during the sequencing reactions, and these beads are then deposited onto a flow cell slide. The amplification starts as a template anneals to the P1 primer, then polymerase extends from the P1 primer, followed by complementary sequence extending from paramagnetic bead. The emPCR takes place in a cell-free system (water-in-oil emulsion) (Nakano *et al.*, 2003) to generate multiple copies of a single DNA fragment onto a bead. The aqueous phase contains PCR reagents, including beads coated with one of the PCR primers. Each compartment in the aqueous phase may contain more than one bead, which may give rise to mixed sequences. After the successful amplification and enrichment, the emPCR beads are breaking apart using butanol, and millions of beads bearing amplification products are enriched by chemically cross-linking to an amino-coated glass surface. After breaking apart the emulsion products, beads bearing amplification products are selectively recovered, and they are then immobilized to a solid planar substrate to generate a dense, disordered array.

Sequencing by ligation

The *ligase*-mediated sequencing approach of the SOLiD SGS platforms uses an adapter-ligated fragment library similar to those of the Roche SGS platforms. The sequencing by ligation begins by annealing a primer to the shared adapter sequences on each amplified fragment, and then DNA ligase is provided along with specific fluorescent labelled octamers, whose 4th and 5th bases are encoded by the attached fluorescent group. Each ligation step is followed by fluorescence detection, after which a regeneration step removes bases from the ligated octamer (including the fluorescent group) and concomitantly prepares the extended primer for another round of ligation. The key features in SOLiD ligase-mediated sequencing are as follows:

1. The sequencing reagents for SOLiD SGS platforms include the DNA ligase, a universal primer, a P1 primer, and sequencing probes.
2. The SOLiD sequencing probes are fluorescent octamers representing all possible sequence variants for an oligo of that size (16 probe types representing 1024 unique probe sequences, see below). Because of this, a highly specific ligation reaction is performed only when the correct base-paired probe anneals adjacent to the primer. In this way, an extended duplex is formed.
3. The fluorescent probes also possess blocking groups at their non-ligating termini to ensure that only a single ligation reaction can take place for each matching sequence. Any unextended template molecules are then capped to prevent their participation in further rounds. The fluorescent label is used to identify the probe, and hence the template sequences.
4. At the 3′ end of each octamer (i.e. the ligating end) are nucleotides that are identified by the fluorophore. But, instead of one nucleotide at the 3′ end being coded by one fluorophore, in the SOLiD system it is the two bases at the probes' 3′ termini that determine which fluorophore is used. This type of chemistry is termed 2-base encoding.
5. In addition to the two specific nucleotides, each octamer has three random (*n*) bases followed by three universal nucleotide analogues (i.e. they do not have specific base-pairing requirements) and the fluorophore. There are thus 16 possible probe types within

the library, representing all combinations of the two nucleotides at the 3′ terminus.

6 In each reaction, four different fluorophores are used. That is, in order to identify which bases have just been sequenced, it is necessary to know both the preceding base and the probe/fluorophore combinations.

7 Since the P1 adapter is a defined sequence, the 5′ base of P1 must be sequenced to allow deconvolution of the remaining sequence. The justification for using this version of the chemistry, as opposed to just the single nucleotide at the probe's 3′ end, is that 2-base encoding allows each base in the template to be interrogated twice, which keeps the error rate low.

8 Following ligation of the correct octamer and imaging of the slide, the three nucleotide analogues are cleaved off, along with the fluorophore, leaving the two specific bases and three Ns. This ability to shorten the probe after ligation means that fewer primers are needed to fill in the entire sequence.

9 The cleaved probe is five bases long, and so five primers are required: the first primer reveals the fluorophore corresponding to bases 1 and 2, 6 and 7, 11 and 12, etc. The second primer is one base shorter, allowing the identification of bases 0 and 1, 5 and 6, 10 and 11, and so on. For additional accuracy, it is possible to sequence with an additional primer using a multibase encoding scheme, although this will increase run time and may not be necessary for all sequencing applications.

Three key events of the cycling sequencing by ligation are illustrated in Figs. 1.10–1.12. The first cycling sequencing event (Fig. 1.10) starts with the preparation of single-stranded template of unknown sequence, which is flanked by known sequences using adapters, followed by attaching it to a solid surface, such as 1 μm beads, and annealing a primer to this known region. The sequencing by ligation begins by annealing a primer to the shared adapter sequences on each amplified fragment, and then DNA ligase is provided along with specific fluorescently labelled octamers, whose fourth and fifth bases are encoded by the attached fluorescent group. Each ligation step in cycling sequencing is followed by fluorescence detection, after which a regeneration step removes bases from the ligated octamer (including the fluorescent group) and concomitantly prepares the extended primer for another round of ligation, i.e. the second cycle of ligation reaction (Fig. 1.11). The multiple rounds of ligation were each followed by periodical reset (Fig. 1.12).

Third-generation sequencing platforms

As an alternative to the SGS technologies, commercial companies are also developing third generation sequencing (TGS) platforms using advanced novel approaches of both optical and non-optical sequencing that include the use of scanning tunnelling electron microscope (TEM), fluorescence resonance energy transfer (FRET), single-molecule detection, and protein nanopores. A major advantage of the TGS platforms is that they eliminate the amplification step, and directly sequence single DNA molecules bound to a surface, thus have the potential to subsequently reduce the sequencing costs even more steeply than SGS instruments. For instance, the Pacific Biosciences TGS platforms relied on optical detection of fluorescent events, targeting to increase sequencing speed and throughput. However, based on non-optical approaches, the Ion Torrent TGS platforms utilize the ion-sensitive field effect transistors (ISFETs) to eliminate the need for optical detection of sequencing events. In contrast, the Oxford Nanopore TGS platform measures changes in conductivity across a Nanopore and removes the optics as well as eliminates the need for DNA amplification in their sequencing design. There is another non-optical TEM approach of TGS platforms being developed by Halcyon Molecular and ZS Genetics. It is not yet fully commercialized in the global market. In this chapter, an overview of the four leading TGS platforms are described, namely Life Technologies' Ion Torrent, Pacific Biosciences' Single Molecule Sequencing (SMS) technology, Helicos Biosciences' Single Molecule Real Time (SMRT) technology and Oxford Nanopore.

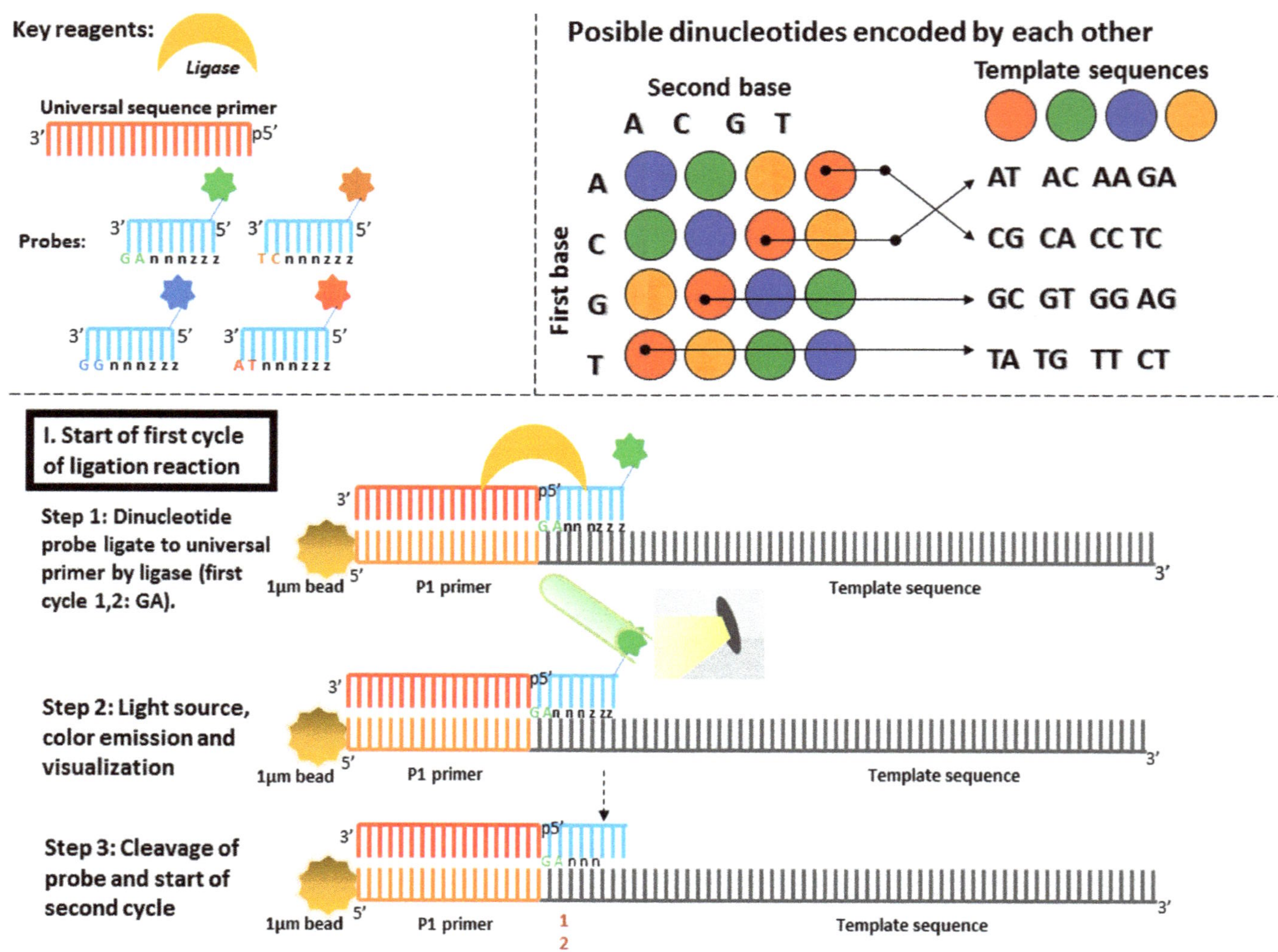

Figure 1.10 Sequencing by Ligation in SOLiD SGS platforms: First cycle of ligation reaction.

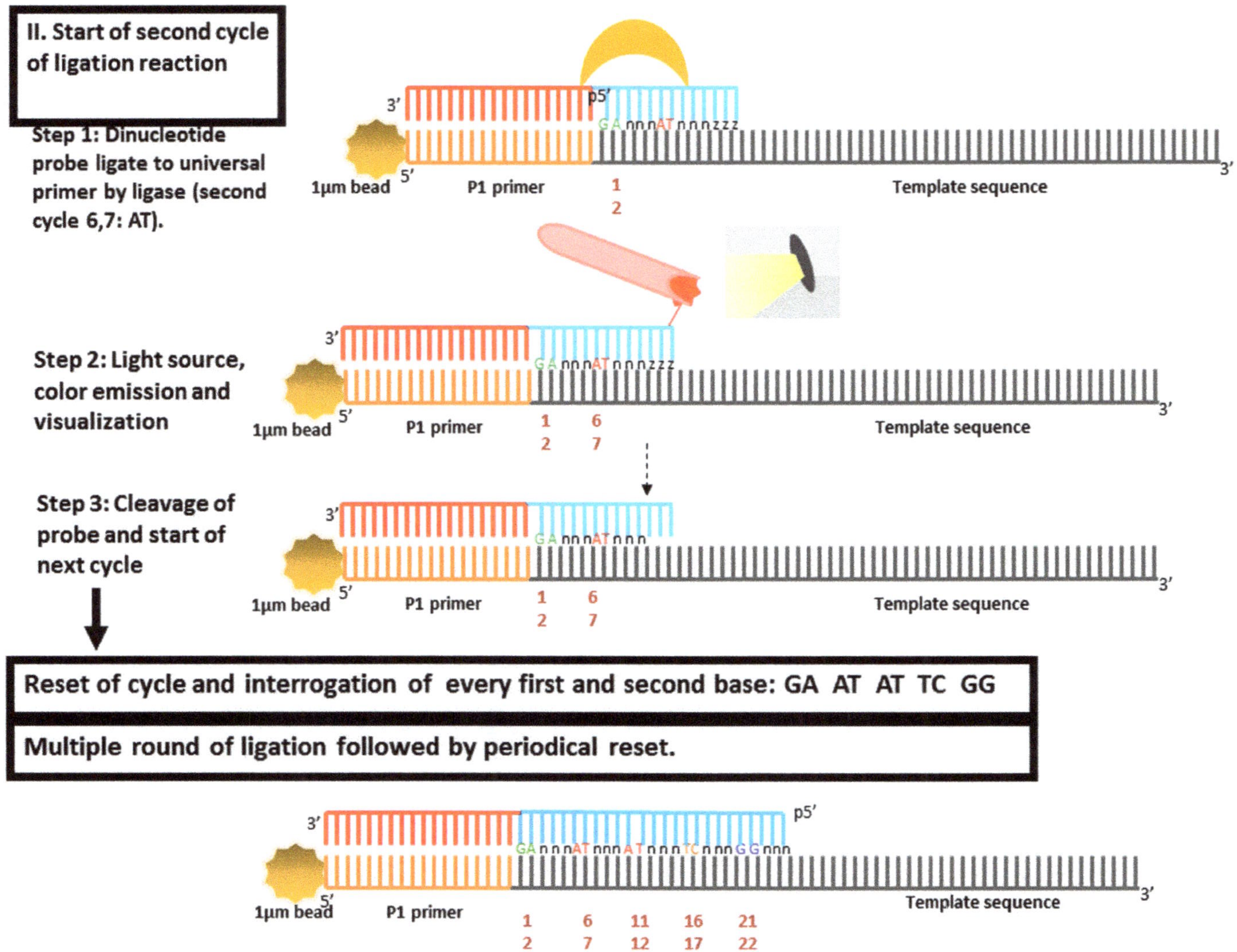

Figure 1.11 Sequencing-by-ligation in SOLiD SGS platforms: Second cycle of ligation reaction.

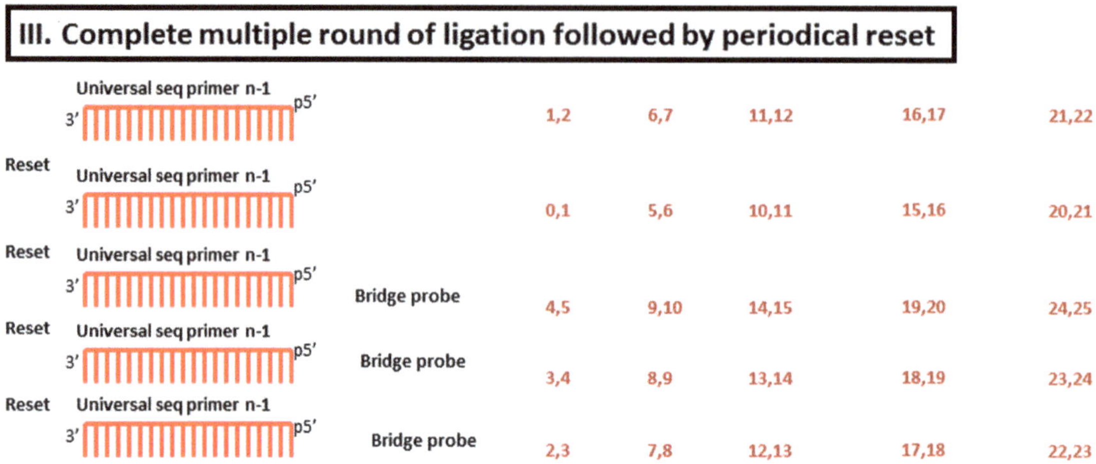

Figure 1.12 Sequencing-by-ligation in SOLiD SGS platforms: Complete multiple round of ligation followed by periodical reset.

Ion Torrent semiconductor TGS platform

Available instruments

At the end of 2010, Ion Torrent (http://www.iontorrent.com/) introduced its first Ion Personal Genome Machine (PGM™). The successor to Ion PGM is called the Ion Proton, released in 2012, and is based on the same chemistry as its predecessor and is expected to have about 60-fold more wells on its Ion Proton II chip. This massive increase in throughput should enable the sequencing of a human genome in a few hours on the machine. The highlights and important features of the Ion Torrent TGS instruments are presented in Table 1.4.

Sequencing chemistry

The Ion Torrent (Life Technologies) TGS platforms constitute a shift in technology from optical-based sequencing systems which measure fluorescence or luminescence output, to monitoring release of hydrogen ions during DNA synthesis in a semiconductor-sensing device (Devonshire *et al.*, 2013). Ion Torrent is a unique TGS platform that utilizes both nucleic acid Watson's chemistry and proprietary of semiconductor technology –Moore's law (Moore, 1965). The principle of sequencing chemistry is based on the PostLightTM sequencing technology (Fig. 1.13), which relies on electric rather than optical fluorescent signals to obtain sequence data. In other words, Ion Torrent sequencing

Table 1.4 Specification of Ion Torrent TGS instrument

Instrument (Ion Torrent PGM* and Ion Proton II)	Run time (hours)	Read length (bp)	Reads/run (millions)	Yield (Mb/run)	Reagent cost/run[a] ($)	Reagents cost/m[b] ($)	Minimum unit cost (% run)[b] ($)	Error type	Error rate (%)	Chip cost/run ($)
314 Chip	2	100	0.10	>10	500	<50	<750 (100%)	indel	1.71	225
316 Chip	2	>100	1	>100	750	<7.5	~$1000 (100)	Indel	>1	425
318 Chip	2	>100	4–8	>1000	~925	~0.93	~$1200 (100%)	indel	>1	625

[a] Includes all stages of sample preparation for a single sample (i.e. library preparation through sequencing).
[b] Typical full cost (i.e. including labour, service contract, etc.) of the smallest generally available unit of purchase at an academic core laboratory provider for the longest available read (and percentage of reads relative to a full run, rounded to the nearest whole percentage).
* The cost of Ion Torrent PGM is $80490 (Information based on company sources alone).

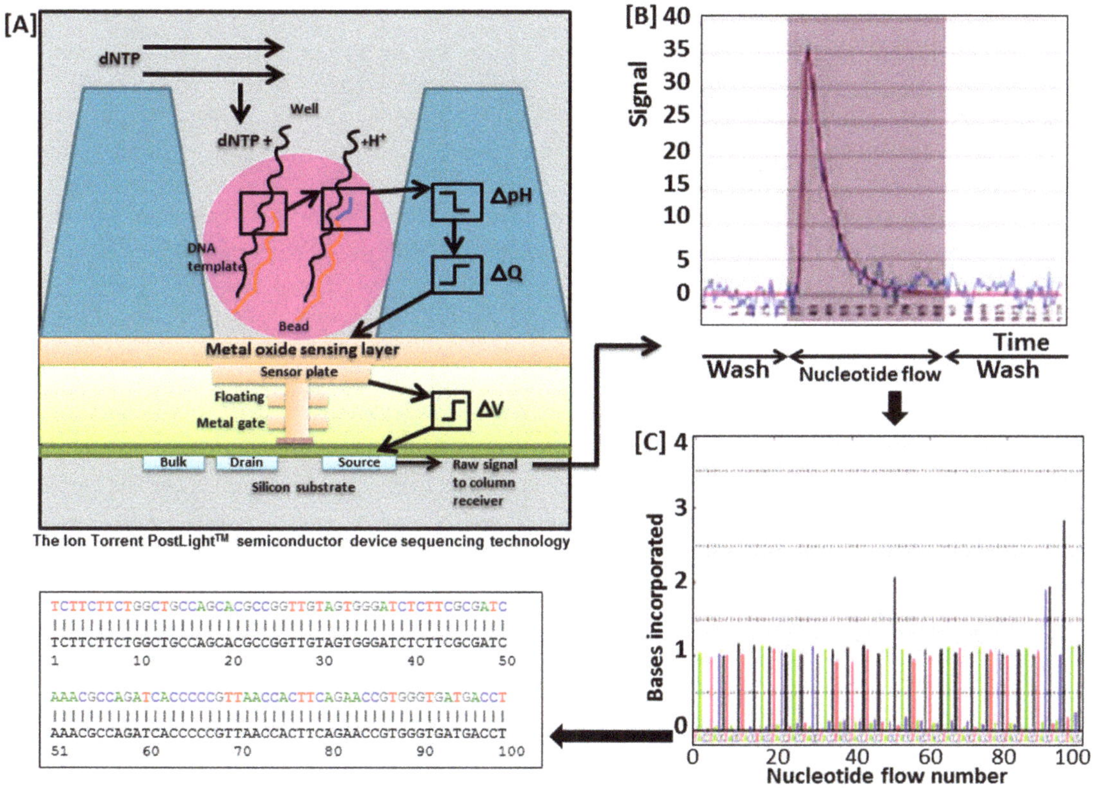

Figure 1.13 Sequencing chemistry of Ion Torrent semiconductor-sensing device.

chemistry is the only commercially available sequencing chemistry that uses entirely natural nucleotides and coupled to an integrated semiconductor device enabling non-optical genome sequencing (Rothberg *et al.*, 2011). Therefore, the major advantage of this TGF platform is that it avoids the need for modified nucleotides, expensive optics and the time-consuming acquisition and processing of large amounts of fluorescent image data. Thus, the Ion Torrent TGS platforms offer a cheaper, space-efficient and more rapid DNA sequencing. The Ion Torrent TGS uses a semiconductor pH sensor device by configuration of sequencing chemistry with templated DNA bead (pink sphere) deposited in a well above the CMOS pHFET sensor (Fig. 1.17 A). In the sequencing events, the single-stranded clonally templated bead is primed for extension, and bound with polymerase, with specific trial dNTP flows into the well and H^+ (proton) is released if it is incorporated as the next base in this synchronized synthesis reaction. This resulting H^+ release (Delta pH) is converted into a +charge build-up (ΔQ) on the sensing layer, which then through metal pillar, transmits as a voltage change (ΔV) at the transistor gate. This in turn gates the transistor current, and the time-accumulated current signal is sent off-chip, in the form of a voltage signal, via a chip-wide polling process (Fig. 1.13A). These raw pH signals are sampled from the well at high frequency (blue curve), at 100 Hz, and then fit to a model that extracts the net signal (pink curve) (Fig. 1.13B). The net incorporation signals for each flow of each well become the primary input, reflecting no incorporation, 1-mer incorporation, 2-mer incorporation, etc. for each trial flow. These raw data serve as input for the base-calling algorithm that converts the full series of net incorporation signals for that well, over hundreds of trial nucleotide flows, into one corresponding sequence read. (Fig. 1.13C).

General description

In 2010, Life Technologies and Ion Torrent developed the Ion Personal Genome Machine (PGM), which represents an affordable and rapid bench-top system designed for small projects. The Ion Torrent PGM system harbours an array of semiconductor chips capable of sensing minor changes in pH and detecting nucleotide incorporation events by the release of a hydrogen ion from natural nucleotides. Thus, the Ion Torrent system does not require any special enzymes or labelled nucleotides and takes advantage of the advances made in the semiconductor technology and component miniaturization. The Ion Torrent workflow process is very similar to that of the Roche SGS platform developed earlier by Rothberg and co-workers (Margulies *et al.*, 2005), and utilizes of the same library preparation steps, with an adapter-ligated fragment library, which is clonally amplified on beads using emulsion PCR, followed by a DNA polymerase that incorporates nucleotides on a primed DNA template with a predetermined flow of nucleotides. However, different from the Roche SGS platform, this TGS platform shifted from detecting the PPi to the other by-product of nucleotide incorporation, the hydrogen ion (Fig. 1.13). The enriched beads from the emulsion PCR are deposited in small wells that are equipped with a sensor capable of detecting free protons on an ion chip. As the DNA polymerase adds nucleotides to the DNA template, hydrogen ions are released and the resulting free proton shift is detected by the sensor and transformed to an electric signal. The instrument has no optical components, and is comprised primarily of an electronic reader board to interface with the chip, a microprocessor for signal processing, and a fluidics system to control the flow of reagents over the chip. The signal processing software converts the raw data into measurements of incorporation in each well for each successive nucleotide flow. After bases are called, each read is passed through a filter to exclude low-accuracy reads and per-base quality values are predicted. The ion chip is based on the most widely used technology to construct integrated circuits, facilitating low-cost and large-scale production and should confer excellent

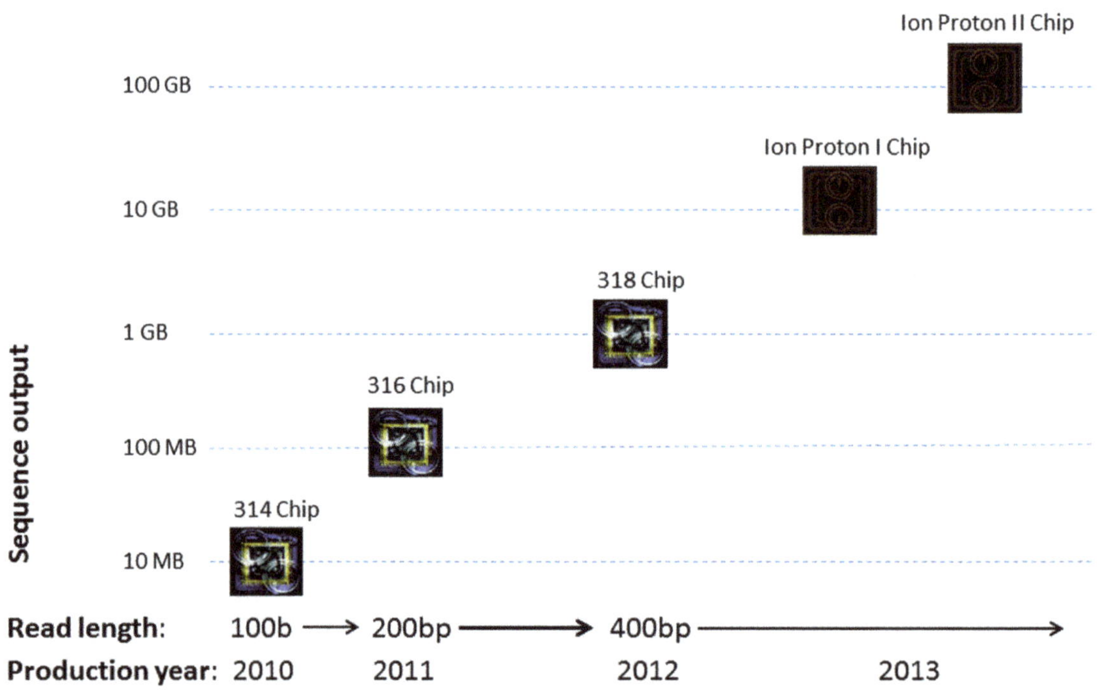

Figure 1.14 Ion Torrent yield trajectory.

scalability. A unique feature of this TGS platform is that the semiconductor device ion chip is itself a machine. Among sequencing platforms, the Ion Torrent system has a fast sequencing run time of just 2 hours. At present, Ion Torrent offers the one-time-use Ion 314 sequencing chip, second generation the Ion 316 sequencing chip and the third generation Ion 318 sequencing chip. The Ion 314 chip comprises 1.2 million microwells that generates roughly 10 Mb of sequence information with average read lengths on the order of 100 bases. However, the Ion 316 chip contains 6.2 million microwells and the Ion 318 chip has 11.1 million microwells and they are capable of producing 100 Mb and 1 Gb of sequencing data with average read lengths of 200–400 bases (Fig. 1.14).

Single molecule sequencing (SMS) TGS platforms using fluorescence-based image detection

Available instruments

In this section, two SMS TGS technologies: the true single molecule sequencing (tSMS) technology developed by Pacific Biosciences and Single Molecule Real Time (SMRT) technology developed by Helicos Biosciences. Another single molecular sequencing TGS platform, i.e. Oxford Nanopore, is discussed separately in the last section of this chapter. The highlights and important features of PacBio RS (Pacific Biosciences) and HeliScope™ (Helicos Biosciences) instruments are presented in Table 1.5.

Sequencing chemistry

Helicos tSMS technology

Helicos Biosciences introduced the first single-molecule DNA sequencer, the HeliScope™ sequencer system, in 2008 (Efcavitch and Thompson, 2010). The platform uses the single-molecule sequencing-by-synthesis approach with immobilized DNA on a planar surface that detects template extension using Cy3 and Cy5 labels attached to the sequencing primer and the incoming nucleotides, respectively. In this approach, the poly(A) tailed template molecules are captured by hybridization to surface-tethered poly-T oligomers to yield a disordered array of primed single molecule sequencing templates. Templates are labelled with Cy3, such that imaging can identify the subset of array coordinates where a sequencing read is expected. Each cycle consists of the polymerase-driven incorporation of a single species of fluorescently labelled nucleotide at a subset of templates, followed by fluorescence imaging of the full array and chemical cleavage of the label (Fig. 1.15).

PacBio RS SMAT technology

This technology is based on the four-colour real-time sequencing chemistry. It harnesses the natural process of DNA replication by DNA polymerase, which attaches itself to a strand of DNA to be replicated, examines the individual base at the point it is attached, and then determines which of the four-colour nucleotides (fluorescent labelled) is required to replicate that individual base. After

Table 1.5 Specification of SMS TGS instruments

Instrument	Run time (hours)	Read length (bp)	Reads/ run (millions)	Yield (Gb/ run)	Reagent cost/ run[a] ($)	Reagents cost/mb ($)	Minimum unit cost (% run) ($)[b]	Error type	Error rate (%)	Purchase cost ($)
HeliScope™	N/A	35	800	28	N/A	N/A	1100 (2%)	deletions	15	N/A
PacBio RS	0.5 –2	860–110	0.01	5–10	110–900	11–180	N/A	indels	12.86	695 000

[a]Includes all stages of sample preparation for a single sample (i.e. library preparation through sequencing).
[b]Typical full cost (i.e. including labour, service contract, etc.) of the smallest generally available unit of purchase at an academic core laboratory provider for the longest available read (and percentage of reads relative to a full run, rounded to the nearest whole percentage).

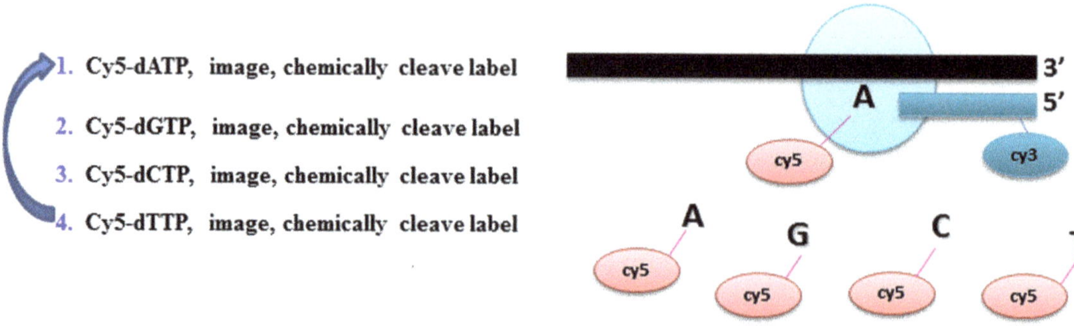

Figure 1.15 Single molecule sequencing-by-synthesis chemistry in Helicos tSMS TGS platform.

determining which nucleotide is required, the polymerase incorporates that nucleotide into the growing strand. After incorporation, the enzyme advances to the next base to be replicated and the process is repeated. In this way, SMRT technology enables the observation of DNA synthesis as it occurs in real time (Fig. 1.16). The technology introduced several interesting novel features, including immobilization of the DNA polymerase instead of the DNA template, real-time monitoring of the synthesis process and potential read lengths of several kilobases of DNA (Eid *et al.*, 2009). Other important features of this technology relate to the design of a nano-structure, including the zero-mode waveguide (ZMW) for real-time observation of DNA polymerization (Levene *et al.*, 2003). Because of this, it reduces the observation volume, therefore reducing the number of stray fluorescently labelled molecules that enter the detection layer for a given period. A ZMW is a hole, tens of nanometres in diameter, fabricated in a 100nm metal film deposited on a glass substrate. These ZMW detectors address the dilemma that DNA polymerases perform optimally when fluorescently labelled nucleotides are present in the micromolar concentration range, whereas most single-molecule detection methods perform optimally when fluorescent species are in the pico- to nanomolar concentration range. The residence time of phospholinked nucleotides in the active site is governed by the rate of catalysis and is usually on the millisecond scale. This corresponds to a recorded fluorescence pulse, because only the bound, dye-labelled nucleotide occupies the ZMW detection zone on this timescale. The released, dye-labelled pentaphosphate by-product quickly diffuses away, dropping the fluorescence signal to background levels. Translocation of the template marks the interphase period before binding and incorporation of the next incoming phospholinked nucleotide.

Key featuring point of tSMS and SMRT technologies

Both tSMS and SMRT technologies are carried out by the sequencing-by-synthesis approach using laser excitation to generate a fluorescent signal from labelled nucleotides, which is then detected using a camera. However, they differ in the following key steps.

1. In tSMS (Ozsolak and Milos, 2011b) technology, single nucleotides, each with a fluorescent dye attached to the base, are sequentially added, whereas, in SMRT technology (Korlach *et al.*, 2010), four different nucleotides, each with a different colour dye attached to the phosphates, are continuously added.
2. Background fluorescence is minimized differently in both technologies. The tSMS technology uses total internal reflectance fluorescence (TIRF) to create a narrow evanescent field of light in which the intensity of the light decays exponentially away from the glass surface and only dyes that are in the TIRF evanescent field can fluoresce. However, the SMRT technology uses a zero mode waveguide (ZMW), which limits illumination to a narrow region near the bottom of the

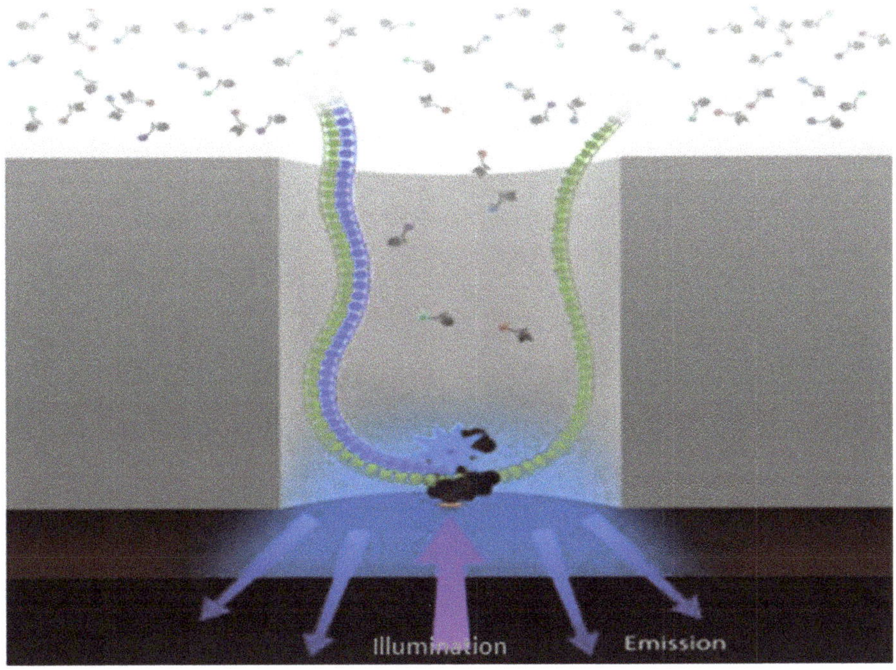

Figure 1.16 Single molecule sequencing-by-synthesis chemistry in Pacific SMTP TGS platform.

well containing the polymerase and only dyes near the opening of the ZMW can fluoresce.

3 In tSMS technology, the immobilization of DNA takes place for viewing over time by a surface-attached sequencing primer, whereas, SMRT performed the DNA immobilization by interaction with a surface-bound polymerase.

4 In tSMS technology, the polymerase is replaced after every cycle of nucleotide addition, in contrast to SMRT where the polymerase cannot be replaced.

General description

Helicos tSMS technology (Fig. 1.17)

Helicos introduced the first commercial single molecule DNA sequencing system: The HeliScope™ in 2007, which is principally based on the 'true single molecule sequencing' (tSMS) of DNA or RNA molecules captured on its flow-cell surface (Braslavsky *et al.*, 2003). The general workflow starts with the constructed single-stranded DNA library that are disorderly arrayed on a flat substrate without any amplification. The nucleic acid fragments are hybridized to primers covalently anchored in random positions on a glass cover slip in a flow cell. The primer, polymerase enzyme and labelled nucleotides are added to the glass support. The next base incorporated into the synthesized strand (array that have undergone template-directed base extension) is determined by analysis of the emitted light signals, which are recorded with a CCD camera. After washing, fluorescent labels on the extended strands are chemically removed, and another cycle of base extension repeats. This system is capable of analysing millions of single-stranded DNA fragments simultaneously, resulting in sequence throughput in the gigabase range. However, one of the key challenges this technology faces is the raw sequencing accuracy due to the difficulty with detecting single molecule event.

PacBio RS SMRT technology (Fig. 1.18)

The PacBio RS system uses 'single molecule real-time detection' (SMRT) which detects the fluorescence of a labelled nucleotide as it is incorporated into the growing DNA strand (Eid *et al.*,

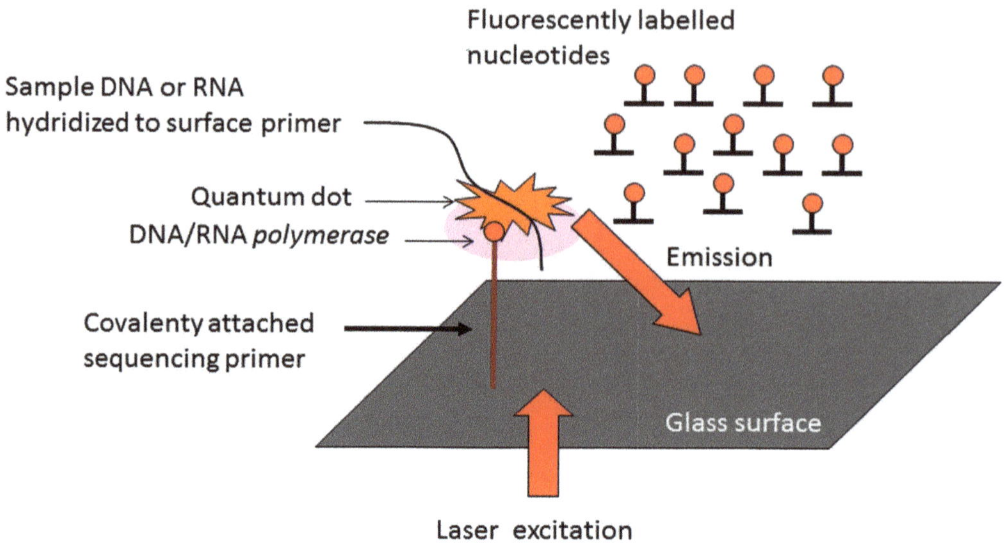

Figure 1.17 An overview of Helicos tSMS technology.

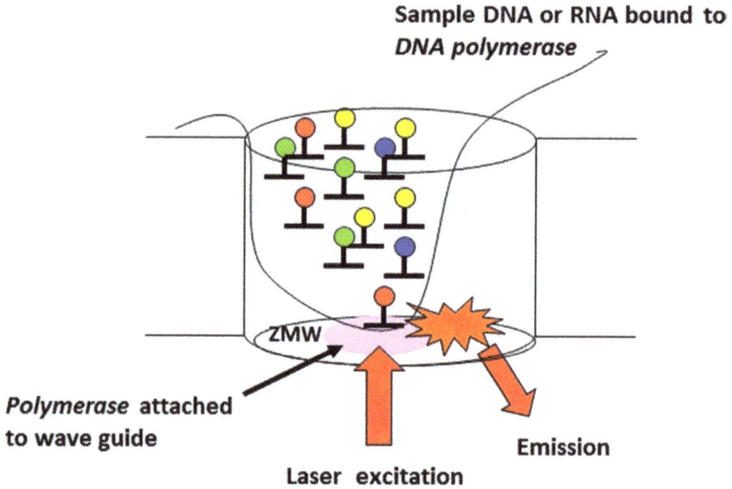

Figure 1.18 An overview of PacBio RS SMRT technology.

2009). The unique feature of this technology is a dense array of zero-mode waveguide (ZMW) nanostructures that allow for optical interrogation of single fluorescent molecules. Fluorescence from a single DNA polymerase molecule is measured per perforation in a metal sheet containing 75,000 such perforations. The fluorescent label is initially attached to the dNTP phosphate group but is cleaved during nucleotide incorporation. Therefore there is no need to reverse terminators and, as each nucleotide is separately labelled, no need to cyclically alternate the availability of nucleotides (Devonshire *et al.*, 2013). The general workflow of PacBio RS platform is as follows.

Preparation of the sequencing library

SMRTbell is the default method for preparing sequencing libraries for PacBio RS in order to get high-accuracy variant detection (Travers *et al.*, 2010). This includes the preparation of randomly fragmented DNA, end-repaired, addition of 3′ adenine to the fragmented genomic DNA, and ligation of an adapter with a T overhang. Single DNA oligonucleotide, which forms an intramolecular hairpin structure, is used as the adapter. The SMRTbell DNA template is structurally a linear molecule but the bubble adapters create a topologically circular molecule.

The SMRT cell

The SMRT cell includes a patterned array of zero-mode waveguides (ZMWs) (Korlach *et al.*, 2008a), which is prepared for *polymerase* immobilization by coating the surface with streptavidin. The preparation of the sequencing reaction requires incubating a biotinylated Phi29 DNA polymerase with primed SMRTbell DNA templates that includes an emulsion PCR amplification step.

Processive DNA sequencing by synthesis

In sequencing reaction, the tethered polymerase incorporates nucleotides with individually phospholinked fluorophores, with each fluorophore corresponding to a specific base, to the growing DNA chain (Korlach *et al.*, 2008b). During the initiation of a base incorporation event, the fluorescent nucleotide is brought into the polymerase's active site and into proximity of the ZMW glass surface. At the bottom of the ZMW, a high-resolution camera records the fluorescence of the nucleotide being incorporated. During the incorporation reaction a phosphate-coupled fluorophore is released from the nucleotide and that dissociation diminishes the fluorescent signal. While the polymerase synthesizes a copy of the template strand, incorporation events of successive nucleotides are recorded. During this sequencing assay, each time a fluorescently labelled base is grabbed on by the polymerase, it brings fluorophore to the detection volume, creating a burst of fluorescent light. If the nucleotide is complementary to the template strand, it will go through a time-consuming synthesis process, and therefore, stay in the detection volume longer until the fluorescent moiety is released as part of pyrophosphate. The colour-coded fluorescence burst and its duration reveal the identity of the complementary base on template DNA. Following the progressive bursts of fluorescence at each waveguide in real time, sequences of template DNA can be rapidly determined. Thus, this technology has a great potential to achieve high speed with long read length. However, a major drawback of error stemmed from real-time single-molecule detection, resulting in low accuracy reads as with other SMS-based TGS platforms.

Single molecule sequencing (SMS) TGS platforms using electronic-based image detection

The SMS technologies have become less expensive and equally capable of producing vast quantities of data. Previously discussed tSMS and SMRT technologies are limited by their reliance on a fluorescent signal, as neither approaches lend themselves particularly well to SMS. This is because they are easy to miss synthesis events and that fluorescence-based imaging requires expensive bespoke nucleotides. The advantages of developing a SMS system that converts base information directly into an electronic signal are thus evident. The currently developed nanopore-based sequencing is based on the electronic detection and is capable of potentially deliver real-time SMS. The initial idea of sequencing DNA by pulling single strands through a nanopore was proposed in the mid-1990s, (Kasianowicz *et al.*, 1996) as an approach that would not only bypass the need for fluorescent detection but also be compatible with SMS. The nanopore technology is based on the principle that if an electrical potential is applied to either side of a nanopore, molecules that pass through the nanopore perturb the current in a characteristic way, and this allows their identification. Currently, the specifications of commercial nanopore-based sequencer instruments are not available, although several companies are working to introduce them in the near future.

Sequencing chemistry of nanopore SMS TGS platform

Current available Oxford Nanopore platform (www.oxfordnanopore.com) sequencing chemistry uses protein pores held in a lipid bilayer. The bilayer is formed across a microwell, with electrodes placed on either side of the bilayer (Fig. 1.19). One protein pore is introduced into the

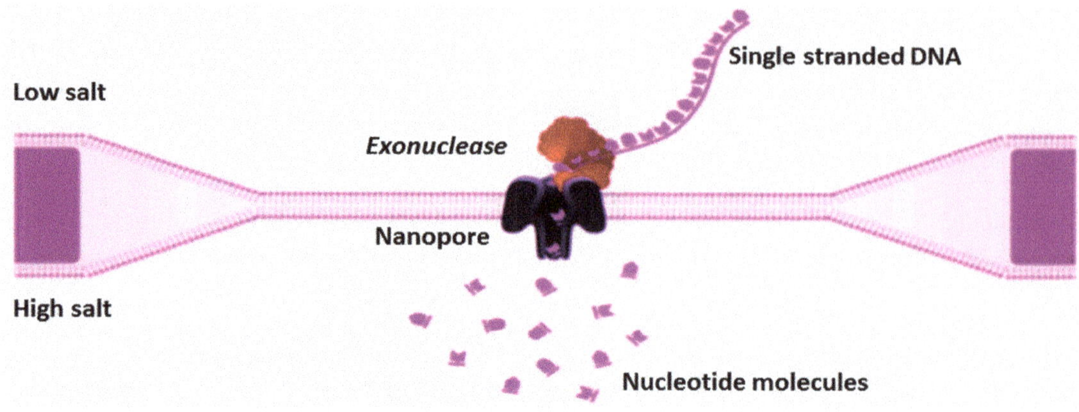

Figure 1.19 Sequencing chemistry of Oxford Nanopore technology.

bilayer, creating a single pore in each microwell. The lipid bilayer has a high electrical resistance, and when an electrical potential is applied across this membrane, a current flows only through the nanopore, carried by the ions in salt solutions on both sides of the bilayer. In general, the nanopores consist of an orifice slightly larger than the width a double-stranded DNA molecule, which is 4 nm, where DNA is threaded through the pore, and the chemical differences of each base would result in detectably altered current flow through the pore. Moreover, the nanopores could also be designed to measure and read the tunnelling current across the pore as bases. Principally, it uses the scanning tunnelling electron microscope (STEM) that measures alterations of conductivity across a nanopore while a single DNA molecular is passing through. The amount of current that can pass through the nanopore at any given moment varies depending on the shape, size, and length of the nucleotide blocking the ion flow through the pore. The change in current through the nanopore as the DNA molecule passes through represents a direct reading of the DNA sequence. An exonuclease enzyme is used to cleave individual nucleotide molecules from the DNA, and when coupled to an appropriate detection system these nucleotides could be identified in the correct order (Clarke *et al.*, 2009).

General description

The Nanopore TGS platform is based on a theory that recording the current modulation of nucleic acids passing through a pore could be used to discern the sequence of individual bases within the DNA chain. Generally, it works with two strategies: *exonuclease* sequencing (biological) and strand (solid) sequencing.

The first *exonuclease* sequencing strategy (Stoddart *et al.*, 2009) works with a processive *exonuclease*, which is attached to the top of the pore (*exonuclease*-based nanopore). So, when a DNA sample is added, the template is digested by the *exonuclease*, releasing the cleaved base. This approach employs a modified α-haemolysin protein with an attached *exonuclease*, situated within a synthetic membrane with high electronic resistance. The *exonuclease* cleaves off single nucleotides and feeds them into the pores. Each individual nucleotide can then be detected from their distinct electrical signal as cyclodextrin attached within the α-haemolysin pore (Astier *et al.*, 2006) to act as a binding site for individual DNA bases, allowing more accurate measurement of their passage through the nanopore binding site (Goren *et al.*, 2009). The second strand sequencing strategy (Howorka *et al.*, 2001; Luan *et al.*, 2010) works with a processive enzyme to draw a single template strand through the nanopore in a controlled fashion. This approach is still underdevelopment and is based on the use of nanopores fabricated either mechanically in silicon or other derivatives. Recently, a considerable amount of progress has been made on this strategy (Stoddart *et al.*, 2010; Olasagasti *et al.*, 2010; Wallace *et al.*, 2010).

Concluding remarks

Technology and funding in the field of novel DNA sequencing platforms have been growing at an unprecedented manner. As discussed in this chapter, there has been a growing increase of vastly different approaches to NGS platforms, across all generations of the newer technologies. The SGS and TGS platforms boast considerable benefits of high-throughput, high accuracy, long contiguous read lengths, and long-range mapping that would be needed to understand and unmask the complex genome of all living organisms.

Acknowledgement

Supported and funded by the National Science Centre, Kraków, Poland (Project No. 2012/05/B/NZ2/01629).

References

Astier, Y., Braha, O., and Bayley, H. (2006). Toward single molecule DNA sequencing: direct identification of ribonucleoside and deoxyribonucleoside 5′-monophosphates by using an engineered protein nanopore equipped with a molecular adapter. J. Am. Chem. Soc. *128*, 1705–1710.

Braslavsky, I., Hebert, B., Kartalov, E., and Quake, S.R. (2003). Sequence information can be obtained from single DNA molecules. Proc. Natl. Acad. Sci. U.S.A. *100*, 3960–3964.

Clarke, J., Wu, H.C., Jayasinghe, L., Patel, A., Reid, S., and Bayley, H. (2009). Continuous base identification for single-molecule nanopore DNA sequencing. Nat. Nanotechnol. *4*, 265–270.

Devonshire, A.S., Sanders, R., Wilkes, T.M., Taylor, M.S., Foy, C.A., and Huggett, J.F. (2013). Application of next-generation qPCR and sequencing platforms to mRNA biomarker analysis. Methods *59*, 89–100.

Dressman, D., Yan, H., Traverso, G., Kinzler, K.W., and Vogelstein, B. (2003). Transforming single DNA molecules into fluorescent magnetic particles for detection and enumeration of genetic variations. Proc. Natl. Acad. Sci. USA, *100*, 8817–8822.

Efcavitch, J.W., and Thompson, J.F. (2010). Single-molecule DNA analysis. Annu. Rev. Anal. Chem. (Palo Alto Calif) *3*, 109–128.

Eid, J., Fehr, A., Gray, J., Luong, K., Lyle, J., Otto, G., Peluso, P., Rank, D., Baybayan, P., Bettman, B., *et al.* (2009). Real-time DNA sequencing from single polymerase molecules. Science *323*, 133–138.

Goren, A., Ozsolak, F., Shoresh, N., Ku, M., Adli, M., Hart, C., Gymrek, M., Zuk, O., Regev, A., Milos, P.M., *et al.* (2009). Chromatin profiling by directly sequencing small quantities of immunoprecipitated DNA. Nat. Methods *7*, 47–49.

Howorka, S., Cheley, S., and Bayley, H. (2001). Sequence-specific detection of individual DNA strands using engineered Nanopores. Nat. Biotechnol. *19*, 636–639.

Kasianowicz, J.J., Brandin, E., Branton, D., and Deamer, D.W. (1996). Characterization of individual polynucleotide molecules using a membrane channel. Proc. Natl. Acad. Sci. U.S.A. *93*, 13770–13773.

Korlach, J., Marks, P.J., Cicero, R.L., Gray, J.J., Murphy, D.L., Roitman, D.B., Pham, T.T., Otto, G.A., Foquet, M., and Turner, S.W. (2008a). Selective aluminium passivation for targeted immobilization of single *DNA polymerase* molecules in zero-mode waveguide nanostructures. Proc. Natl. Acad. Sci. U.S.A.*105*, 117611–81.

Korlach, J., Bibillo, A., Wegener, J., Peluso, P., Pham, T.T., Park, I., Clark, S., Otto, G.A., and Turner, S.W. (2008b). Long processive enzymatic DNA synthesis using 100% dye-labeled terminal phosphate-linked nucleotides. Nucleosides Nucleotides Nucleic Acids *27*, 1072–1083.

Korlach, J., Bjornson, K.P., Chaudhuri, B.P., Cicero, Flusberg, R.L. B.A., Gray, J.J., Holden, D., Saxena, R., Wegener, J., and Turner, S.W. (2010). Real-time DNA sequencing from single polymerase molecules. Methods Enzymol. *472*, 431–455.

Levene, M.J., Korlach, J., Turner, S.W., Foquet, M., Craighead, H.G., and Webb, W.W. (2003). Zero-mode waveguides for single-molecule analysis at high concentrations. Science *299*, 682–686.

Luan, B., Peng, H., Polonsky, S., Rossnagel, S., Stolovitzky, G., and Martyna, G. (2010). Base-by-base ratcheting of single stranded DNA through a solid-state nanopore. Phys. Rev. Lett. *104*, 238103.

McKernan, K.J., Peckham, H.E., Costa, G.L., McLaughlin, S.F., Fu, Y., Tsung, E.F., Clouser, C.R., Duncan, C., Ichikawa, J.K., Lee, C.C., *et al.* (2009). Sequence and structural variation in a human genome uncovered by short-read, massively parallel ligation sequencing using two-base encoding. Genome Res. *9*, 1527–1541.

Macevicz, S.C. (1998). DNA sequencing by parallel oligonucleotide extensions. US patent 5750341.

Mardis, E.R. (2008). Next-generation DNA sequencing methods. Annu. Rev. Genomics Hum. Genet. *9*, 387–402.

Mardis, E.R. (2011). A decade's perspective on DNA sequencing technology. Nature *470*, 198–203.

Mardis, E.R. (2012). Genome sequencing and cancer. Curr. Opin. Genet. Dev. *22*, 245–250.

Margulies, M., Egholm, M., Altman, W.E., Attiya, S., Bader, J.S., Bemben, L.A., Berka, J., Braverman, M.S., Chen, Y.J., Chen, Z., *et al.* (2005). Genome sequencing in microfabricated high-density picolitre reactors. Nature *437*, 376–380.

Maxam, A.M., and Gilbert, W. (1977). A new method for sequencing DNA. Proc. Natl. Acad. Sci. U.S.A. *74*, 560–564.

Metzker, M.L. (2010). Sequencing technologies – the next-generation. Nat. Rev. Genet. *11*, 31–46.

Moore, G.E. (1965). Cramming more components onto integrated circuits. Electronics Magazine *38*, 1–4.

Nakano, M., Komatsu, J., Matsuura, S., Takashima, K., Katsura, S., and Mizuno, A. (2003). Single-molecule PCR using water-in-oil emulsion. J. Biotechnol. *102*, 117–124.

Olasagasti, F., Lieberman, K.R., Benner, S., Cherf, G.M., Dahl, J.M., Deamer, D.W., and Akeson, M. (2010). Replication of individual DNA molecules under electronic control using a protein Nanopore. Nat. Nanotechnol. *5*, 798–806.

Ozsolak, F., and Milos, P.M. (2011a). RNA sequencing: advances, challenges and opportunities. Nat. Rev. Genet. *12*, 87–98.

Ozsolak, F., and Milos, P.M. (2011b). Transcriptome profiling using single-molecule direct RNA sequencing. Methods Mol. Biol. *733*, 51–61.

Pareek, C.S., Smoczynski, R., and Tretyn, A. (2011). Sequencing technologies and genome sequencing. J. Appl. Genet. *52*, 413–435.

Ronaghi, M., Uhlen, M., and Nyren, P. (1998). A sequencing method based on real-time pyrophosphate. Science *281*, 363–365.

Ronaghi, M., Uhlen, M., and Nyren, P. (1999). Analyses of secondary structures in DNA by pyrosequencing. Anal. Biochem. *267*, 65–71.

Rothberg, J.M., Hinz, W., Rearick, T.M.D., Schultz, J., Mileski, W., Davey, M., Leamon, J.H., Johnson, K., Milgrew, M.J., Edwards, M., *et al.* (2011). An integrated semiconductor device enabling non-optical genome sequencing. Nature *475*, 348–352.

Rusk, N. (2009). Cheap third-generation sequencing. Nat. Methods *6*, 244–245.

Sanger, F., Nicklen, S., and Coulson, A.R. (1977). DNA sequencing with chain-terminating inhibitors. Proc. Natl. Acad. Sci. U.S.A. *74*, 5463–5467.

Schadt, E., Turner, S., and Kasarskis, A. (2010). A window into third-generation sequencing. Hum. Mol. Genet. *19*, R227–240.

Shendure, J., Porreca, G.J., Reppas, N.B., Lin, X., McCutcheon, J.P., Rosenbaum, A.M., Wang, M.D., Zhang, K., Mitra, R.D., and Church, G.M. (2005). Accurate multiplex polony sequencing of an evolved bacterial genome. Science *309*, 1728–1732.

Stoddart, D., Heron, A.J., Mikhailova, E., Maglia, G., and Bayley, H. (2009). Single-nucleotide discrimination in immobilized DNA oligonucleotides with a biological nanopore. Proc. Natl. Acad. Sci. U.S.A. *106*, 7702–7707.

Stoddart, D., Heron, A.J., Klingelhoefer, J., Mikhailova, E., Maglia, G., and Bayley, H. (2010). Nucleobase recognition in ssDNA at the central constriction of the alpha-hemolysin pore. Nano. Lett. *10*, 3633–3637.

Travers, K.J., Chin, C.-S., Rank, D.R., Eid, J.S., and Turner, S.W. (2010). A flexible and efficient template format for circular consensus sequencing and SNP detection. Nucleic Acids Res. *38*, e159.

Valouev, A., Ichikawa, J., Tonthat, T., Stuart, J., Ranade, S., Peckham, H., Zeng, K., Malek, J.A., Costa, G., McKernan K., *et al.* (2008). A high-resolution, nucleosome position map of C. elegans reveals a lack of universal sequence-dictated positioning. Genome Res. *18*, 1051–1063.

Wallace, E.V., Stoddart, D., Heron, A.J., Mikhailova, E., Maglia, G., Donohoe, T.J., and Bayley, H. (2010). Identification of epigenetic DNA modifications with a protein Nanopore. Chem. Commun. *46*, 8195–8197.

Attomole-level Genomics with Single-molecule Direct DNA, cDNA and RNA Sequencing Technologies

2

Fatih Ozsolak

Abstract

With the introduction of next-generation sequencing (NGS) technologies in 2005, the domination of microarrays in genomics quickly came to an end due to NGS's superior technical performance and cost advantages. By enabling genetic analysis capabilities that were not possible previously, NGS technologies have started to play an integral role in all areas of biomedical research. This chapter outlines the low-quantity DNA/cDNA sequencing capabilities and applications developed with the Helicos single molecule DNA sequencing technology.

Introduction

The era of first generation sequencing dominated by Sanger sequencing (Sanger and Coulson, 1975) came to an end with the introduction of the Roche/454 platform (Margulies *et al.*, 2005) in 2005, followed by the Solexa/Illumina system (Bentley *et al.*, 2008) in 2006 and the Applied Biosystems SOLiD system (Valouev *et al.*, 2008) in 2007. These second generation sequencing (SGS) technologies have enjoyed a warm welcome from the scientific community. With such an enthusiastic customer base and investments from the users and SGS companies in technology and applications development, many novel capabilities have been achieved within a very short timeframe after their launch. SGS applications in biomedicine have primarily been whole/targeted genome sequencing, digital gene expression, DNA methylation sequencing and copy-number variation. However, SGS technologies have expanded to chromatin immunoprecipitation sequencing (ChIP-seq), RNA immunoprecipitation sequencing, whole transcriptome analyses, single cell analyses, nucleic acid structure determination, chromatin conformation analysis and many others (Mardis, 2008; Morozova and Marra, 2008). SGS technologies have also been adapted beyond nucleic acid sequencing and, for instance, to protein-DNA affinity measurements (Nutiu *et al.*, 2011), which may find frequent use in biomedicine in the future. In this chapter, the Single Molecule DNA/cDNA Sequencing (SMDS) and Direct RNA Sequencing (DRS) technologies developed and commercialized by Helicos BioSciences Corporation are outlined. Detailed procedures for the application of SMDS for low-quantity/attomole-level DNA/cDNA sequencing applications, such as for ChIP-seq (Goren *et al.*, 2010) or circulating cell-free blood nucleic acids (van den Oever *et al.*, 2012), are also provided.

Single molecule DNA/cDNA sequencing (SMDS) and direct RNA sequencing (DRS) technologies

The first single molecule second- (or third-, depending on the criteria used) generation sequencer was commercialized by Helicos BioSciences Corporation (Cambridge, MA, USA) in 2007 (Harris *et al.*, 2008). The system integrates recent advances in automated single molecule fluorescent imaging, fluidics technologies and enzymology, and enables SMDS and DRS capabilities. The system is the first and currently the only system that can sequence RNA directly (e.g. without cDNA synthesis) in addition to DNA and cDNA molecules (Ozsolak *et al.*, 2009). The system proved its value by enabling discovery of novel classes of short RNAs (Kapranov *et al.*, 2010), genome-wide quantitative mapping of polyadenylation sites (Ozsolak *et al.*, 2010a), characterization of circulating tumour cells (Yu *et al.*, 2012), detecting fetal genetic abnormalities by sequencing maternal plasma DNA (van den Oever *et al.*, 2012), ancient DNA sequencing (Orlando *et al.*, 2011) and many other biological and technical advances and discoveries. The SMDS and DRS chemistry and operating principles have been described in detail previously in multiple publications (Harris *et al.*, 2008; Ozsolak, 2012; Ozsolak and Milos, 2011). Below is a brief summary from the previous presentations of this technology.

The sequencing system consists of two main components (Harris *et al.*, 2008; Ozsolak, 2012): (i) flow cells, which create an environment for template nucleic acid attachment, sequencing chemistry steps and single molecule fluorescence imaging, and (ii) the HeliScope™ sequencer, which is an integrated fluidics, optics (for automated, fast and high-throughput imaging) and image/data analysis system for introduction of sequencing reagent formulations to the flow cells, taking single molecule images and analysing the images in real-time (e.g. during the sequencing run) to deduce read sequences. Sequencing flow cell surfaces are coated with poly(dT) oligonucleotides covalently attached at their 5′ amine ends to an epoxide-coated ultra-clean glass surface. These oligonucleotides serve two purposes: (i) to capture the 3′ poly(A)-tail containing nucleic acids onto surfaces by hybridization, and (ii) to prime and initiate the sequencing steps. The current requirement for nucleic acid sample preparation for this platform is the presence of a 3′ poly(A) tail >25–30 nucleotide (nt) in length and 'blocked' at its 3′ end against extension by the polymerase used in the sequencing-by-synthesis steps. 3′ polyadenylation and blocking of DNA/cDNA templates is performed using terminal transferase with dATP and dideoxyATP, respectively. *E. coli* or yeast poly(A) polymerases with ATP and 3'deoxyATP are used for RNA 3′ poly(A)-tail addition and blocking. However, for the characterization of RNA species that naturally contain a poly(A) tail for the purposes of gene expression measurements, polyadenylation site mapping and other applications, such poly(A)-tail addition is not required and direct hybridization of poly(A)+ RNAs to surfaces can be performed. It is also possible to change the poly(dT) oligonucleotides on the surface to other nucleotides for the capture and sequencing of other targets (for instance, for targeted selection and sequencing of particular DNAs/RNAs of interest in a single step manner) (Thompson *et al.*, 2012).

After hybridization of templates to the poly(dT) surface primers, to begin sequencing at the unique template region adjacent to the poly(A)-tail, each primer–template pair is 'filled' in with thymidine triphosphate by a polymerase, and then 'locked' in position with A, C and G Virtual Terminator™ (VT) nucleotides. VTs (Bowers *et al.*, 2009) are nucleotide analogues used for sequencing, containing a fluorescent dye as part of a chemically cleavable group that prevents the addition of another nucleotide. After washing away the excess, unincorporated nucleotides, the surface is irradiated with a laser at an angle that allows total internal reflection at the surface. In such a situation, an evanescent field is generated so that only molecules very close to the surface are excited by the laser. This reduces the background level of fluorescence such that single molecules can be detected. After image acquisition across desired number of positions per channel, the locations of templates on the surface are recorded. The liquid in each channel is replaced with a mixture that cleaves the fluorescent dye and virtual terminator group off the incorporated nucleotide, rendering the strands suitable for further VT incorporation. The sequencing-by-synthesis reaction consists of polymerase-driven cyclic addition of the VTs in the C, T, A and G order. Each VT addition is followed by a rinsing, imaging (to locate the templates that incorporated the particular VT), and cleavage cycle. Repeating this cycle many times provides a set of images from which the base incorporations are detected and then used to generate sequence information for each template molecule. The sequencing operation principles described above are common for both SMDS and DRS. However, the components and conditions of chemistry and reaction steps are different between the two. As a result, different sequencing kits are provided for DRS and SMDS, and, at this point, it is not possible to sequence

DNA/cDNA and RNA molecules concurrently in a single sequencing run.

Each SMDS and DRS run is currently performed with 120 VT-incorporation cycles and contains up to 50 independent channels, producing up to 25 million aligned reads ≥ 25 nt in length (up to 55 nt in length, median ~ 34 nt) per channel depending on the user-defined run time (1–12 days) and throughput (e.g. imaging quantity per channel). Longer reads can be achieved by doing more cycles of sequencing, although the quality and the yield/efficiency of the sequencing reaction may decrease with longer runs. Error rates are in the range of 4–5%, dominated by missing base errors (~ 2–3%), while insertion (~ 1%) and substitution (~ 0.4%) errors are lower.

The system's single molecule nature and low input DNA/RNA requirements (400–1000 amol) are particularly advantageous for settings with limited cell/nucleic acid quantities (Goren *et al.*, 2010; Ozsolak *et al.*, 2010b). Its avoidance of extensive manipulation steps (Fig. 2.1) result in better coverage, minimal or no GC bias and higher quantification and representation range for DNA and RNA compared with other SGSs (Fig. 2.2), even in ancient, formalin-fixed paraffin-embedded tissue (FFPET) and other highly degraded samples (Orlando *et al.*, 2011). Helicose's DRS today is the only technology that can avoid well-known reverse transcription and sample manipulation biases for sequencing RNAs. One advantage of DRS is the universality of sample preparation steps for different applications. In other words, unlike RNA sequencing methods that require intermediate cDNA molecules and different cDNA synthesis and sample manipulation steps for short and long RNAs, DRS requires only 3′ polyadenylated templates. Thus, both short and long RNAs can be sequenced together in a single experiment. The system's short read lengths, up to 55 nucleotides (nts), lack of paired read capability and error rate poses challenges in certain applications such as alternative splicing detection, microbiome sequencing and de novo genome sequencing. The system may be altered to enable elevated read lengths and paired reads in theory through minor modifications, but the errors may not be improved significantly. This is largely because the majority of the errors are caused by

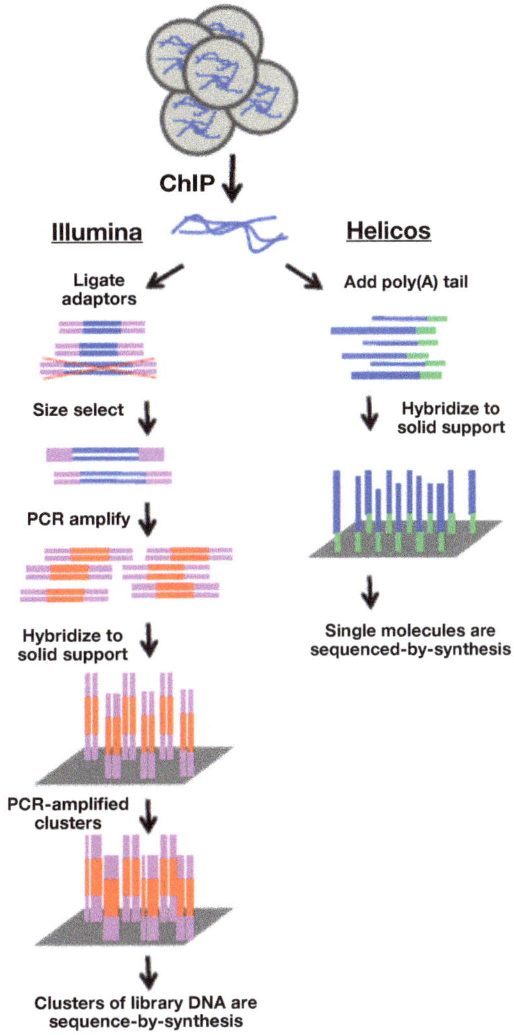

Figure 2.1 ChIP-seq sample preparation and sequencing steps with Illumina and Helicos SMDS technologies. Blue lines: original/unamplified ChIP DNA; green lines: poly(A) tails; pink lines: adaptors; red lines: amplified ChIP DNA; grey rectangles: sequencing flow cell surfaces. The figure was reprinted with permission from the Nature Publishing Group (Goren *et al.*, 2010).

fluorescent dye blinking during the imaging step and there is no apparent remedy for this issue at present. The system is capable of single step target selection and sequencing (Thompson *et al.*, 2012), which provides advantages in targeted sequencing and diagnostics applications focusing on gene sequencing. A modified version of the system has been extended to proteomics (Tessler *et al.*, 2009).

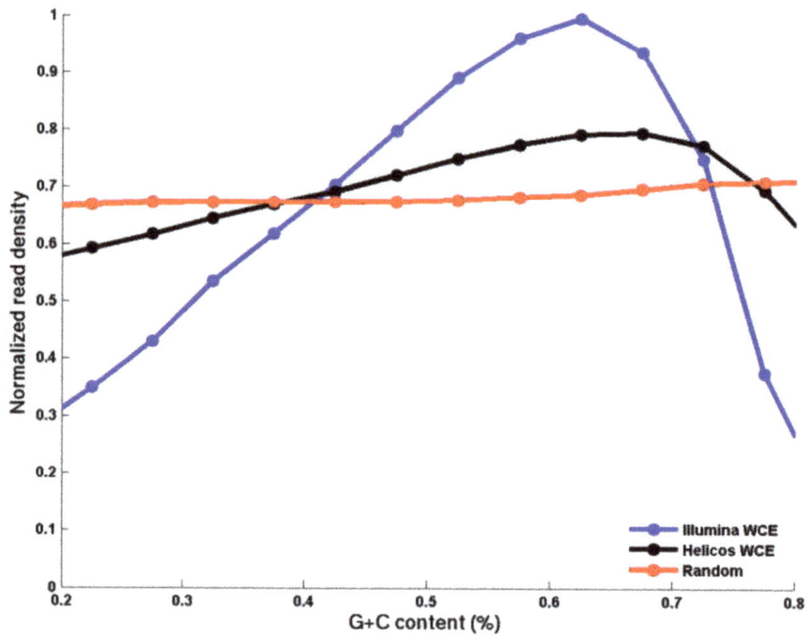

Figure 2.2 Sequencing bias comparison. Normalized read density obtained from un-enriched input control 'whole cell extract' (WCE) samples analysed by Illumina (blue) or Helicos (black) is plotted as a function of GC percentage for 100bp windows. A theoretical 'expected' distribution obtained by computational simulation is shown in red (Random). The Helicos SMDS data show a relatively even distribution across 20–80% GC-content. The figure was reprinted with permission from the Nature Publishing Group (Goren *et al.*, 2010).

Materials

For reagents used, see Table 2.1. For equipment used, see Table 2.2.

Methods

This protocol is for the preparation of low quantity DNA/cDNA samples for SMDS genome re-sequencing, ChIP-seq and other applications (Goren *et al.*, 2010). For simplicity, the protocol assumes that the DNA/cDNA has been fragmented into a desired size range > 50 nucleotides in length. Satisfactory fragmentation results have been obtained with DNA sheared by Covaris, Branson or Misonix sonicators. DNA should be free of RNA contamination. Extensive RNase digestion and clean up with a Reaction Cleanup Kit are highly recommended (Qiagen, catalogue no. 28204). DNA should be accurately quantitated prior to use. The Quant-iT™ PicoGreen dsDNA Reagent Kit with a Nanodrop 3300 Fluorospectrometer is strongly recommended, but standard plate readers may also be used. Molecular biology-grade nuclease-free glycogen or linear acrylamide can be used as carrier during ChIP DNA clean-up/precipitation steps but carrier nucleic acids such as salmon sperm DNA should not be added prior to tailing.

Poly(A) tailing reaction

1. Add the following components in the indicated order: 2 µl of NEB terminal transferase 10× buffer, 2 µl 2.5 mM $CoCl_2$, and 10.8 µl DNA/cDNA in nuclease-free water.
2. Heat the above mix at 95°C for 5 min in a thermocycler for denaturation, followed by rapid cooling on a pre-chilled aluminium block kept in an ice and water slurry (~ 0°C). It is essential to chill the block to 0°C in an ice/water slurry and to cool the sample as quickly as possible to 0°C to minimize re-annealing of the denatured DNA.
3. Mix the following reagents on ice and add to the 14.8 µl denatured DNA from step 2, to a total volume of 20 µl for each sample: 1 µl of

Table 2.1 Reagents

Reagent	Recommended Vendor and Catalogue #
Terminal transferase kit	New England Biolabs M0315
dATP	Roche 11277049001
Biotin-ddATP	Perkin Elmer NEL548001
Non-poly(A) CARRIER OLIGONUCLEOTIDE	IDT
Bovine serum albumin	NEB B9001S
Nuclease-free water	Ambion AM9932
Quant-iT™ PicoGreen dsDNA reagent	Invitrogen P11495

Table 2.2 Equipment

Equipment	Recommended Vendor and Model
Thermal cycler	Bio-Rad DNA Engine Tetrad® 2
Refrigerated microcentrifuge	Eppendorf 5810 R
Aluminium blocks	VWR 13259–260
Nanodrop 3300	Thermo Fisher Scientific

NEB terminal transferase (diluted 4-fold to 5 U/µl using 1× NEB terminal transferase buffer), 4 µl of 50 µM dATP and 0.2 µl of NEB BSA.

4 Place the tubes in a thermocycler and incubate at 37°C for 1 h, followed by 70°C for 10 min to inactivate the enzyme.

3′ End blocking reaction

1 Denature the 20 µl polyadenylation reaction from step 4 from 'poly(A) tailing reaction' above at 95°C for 5 min in a thermocycler, followed by rapid cooling on a pre-chilled aluminium block kept in an ice/water slurry (~0°C).

2 For each sample, add the following blocking mastermix to the denatured DNA from step 1, to a total volume of 30 µl: 1 µl NEB terminal transferase 10× buffer, 1 µl 2.5 mM $CoCl_2$, 1 µl NEB terminal transferase (diluted 4-fold to 5 U/µl using 1× NEB terminal transferase buffer), 0.5 µl 200 µM Biotin-ddATP, and 6.5 µl nuclease-free water.

3 Place the tubes in a thermocycler and incubate at 37°C for 1 h, followed by 70°C for 20 min.

4 Add 2 pmol of the carrier oligonucleotide to the heat-inactivated 30 µl terminal transferase reaction from step 3. At this point, the sample may be stored at −20°C or below for future use or hybridized to a sequencing flow cell directly. The carrier oligonucleotide is added to the sample after the completion of the poly(A) tailing and 3′ blocking steps to minimize template loss during the sample loading steps. It does not contain a poly(A) tail and therefore does not hybridize to flow cell surfaces. The sequence and length (preferably 50–80mer) of the oligonucleotide can be chosen by the user with the only sequence constraint being that it should not interfere with a poly(A):poly-T hybridization. 3″ blocking is preferred to ensure that the polyA tail is not sequenced. The sequence 5′-TCACTATTGTTGAGAAC GTTGGCCTATAGTGAGTCGTATTACGC GCGGT[ddC]-3′ has been used successfully as the Carrier Oligonucleotide but many others can be used.

Flow cell hybridization and SMDS

The detailed procedures for sample loader usage, flow cell rehydration, sample hybridization buffer, fill and lock steps, software and sequencer usage are regularly updated and the latest versions should be obtained from the manufacturer. This section is intended to describe several relevant details for the purposes of SMDS.

Hybridization of samples to flow cell channels is performed in 7–100 μl volume. The samples are mixed 50:50 with 2 hybridization buffer provided in the SMDS Kit. In general, 0.5–2 fmol of DNA/cDNA material is required to optimally load each sequencing channel, although lower quantities can be used for less than optimal aligned read yields per channel. Following hybridization, the template molecules are filled and locked and the flow cells are moved to the HeliScope. The manufacturer-supplied sequencing scripts offer the user the opportunity to run one or two flow cells per run, to define the desired number and location of channels to be sequenced, and to request the desired imaging quantity (e.g. throughput) per channel (the system currently allows up to 1400 fields of view per channel, with up to 25 million aligned reads ≥ 25 nt in length per channel).

Data analysis

The various programs and pipelines for the filtering, alignment and downstream analyses of the data, and the corresponding user manuals can be downloaded freely at http://open.helicosbio.com. Briefly, an initial filtering step is performed on the raw reads before initiating their alignment to reference sequences. This filtering step involves the following read selection steps:

1. SMDS generates reads between 6 nts and 55 nts in length. Depending on the experimental goals, the reference sequence complexity and size, a user-defined minimum read length cut off is employed to remove short reads that cannot be aligned reliably to save computing time and power. ≥ 25 nt cut-off is routinely used for alignment to human and mouse genomes. However, for smaller and less-complex genomes, reads as short as 15 nt can be used.
2. Any 5′ polyT stretches in the reads are trimmed. The likely cause of such T homopolymeric stretches is incomplete fill with dTTP, leading to sequencing initiation within the poly(A) tail. Therefore, 5′ poly-T trimming is preferred to minimize potential misalignment events.
3. Because of flow cell surface imaging errors, artificial reads that have a repetition of the VT-addition order sequence (CTAG) may appear. Such reads are eliminated during the filtering step.

Given that the majority of sequencing errors are due to indels, an aligner that is tolerant to these types of errors should be employed. We highly recommend the use of the indexDPgenomic aligner (Giladi *et al.*, 2010) and downstream data analysis, genotyping and quantification tools freely available at http://open.helicosbio.com. While multiple aligners are available and can align SMDS reads, including Mosaik (http://code.google.com/p/mosaik-aligner/) and SHRiMP (Rumble *et al.*, 2009), the use of these aligners may result in a reduction in actual aligned reads due to their reduced ability to deal effectively with indels.

SMDS ChIP-seq and other application notes

The most robust performance with this protocol and SMDS is generally obtained with 6–9 ng of DNA with an average size of 400–500 nucleotides but 1 ng or less can be used successfully. This protocol has been successfully used for sequencing of fetal DNA from 200–300 pg of maternal cell-free blood nucleic acids (van den Oever *et al.*, 2012). Because sequencing efficiency depends on the number of free 3′ hydroxyls that can be tailed, molarity must be estimated for optimal performance. If the average DNA size is shorter, correspondingly less mass of DNA should be used. If more than 18 ng of DNA is available, only a fraction of it should be used as too much DNA can result in sample overloading and reduced yield. Since the yields of ChIP DNA experiments vary greatly depending on multiple factors (antibody specificity, antibody affinity, ChIP procedures, etc.), ChIP DNA quantity should be determined prior to the polyA tailing reaction with the Quant-iT™ PicoGreen dsDNA Reagent Kit. Depending on the hybridization mixture volume used, the processed ChIP material is sufficient for loading 1–15 channels of a flow cell. Yield is typically 7–12 M aligned reads per channel. End-repair of ChIP DNA may be performed, but has a variable and marginal impact on yield for most samples. Nevertheless, end-repair must be performed on ChIP DNA fragmented using micrococcal nuclease treatment, as this enzyme leaves 3′ phosphate groups that prevent tailing with terminal transferase, or fragmentation methods leaving high fraction of untailable/blocked 3′ ends.

Acknowledgements

The author thanks all past and previous Helicos employees for their help and contribution to the development of the technology and the work described here. This work was partly supported by NHGRI grants R01 HG005230 and R44 HG005279.

References

Bentley, D.R., Balasubramanian, S., Swerdlow, H.P., Smith, G.P., Milton, J., Brown, C.G., Hall, K.P., Evers, D.J., Barnes, C.L., Bignell, H.R., *et al.* (2008). Accurate whole human genome sequencing using reversible terminator chemistry. Nature *456*, 53–59.

Bowers, J., Mitchell, J., Beer, E., Buzby, P.R., Causey, M., Efcavitch, J.W., Jarosz, M., Krzymanska-Olejnik, E., Kung, L., Lipson, D., *et al.* (2009). Virtual terminator nucleotides for next-generation DNA sequencing. Nat. Methods *6*, 593–595.

Giladi, E., Healy, J., Myers, G., Hart, C., Kapranov, P., Lipson, D., Roels, S., Thayer, E., and Letovsky, S. (2010). Error tolerant indexing and alignment of short reads with covering template families. J. Comput. Biol. *17*, 1397–1411.

Goren, A., Ozsolak, F., Shoresh, N., Ku, M., Adli, M., Hart, C., Gymrek, M., Zuk, O., Regev, A., Milos, P.M., *et al.* (2010). Chromatin profiling by directly sequencing small quantities of immunoprecipitated DNA. Nat. Methods *7*, 47–49.

Harris, T.D., Buzby, P.R., Babcock, H., Beer, E., Bowers, J., Braslavsky, I., Causey, M., Colonell, J., Dimeo, J., Efcavitch, J.W., *et al.* (2008). Single-molecule DNA sequencing of a viral genome. Science *320*, 106–109.

Kapranov, P., Ozsolak, F., Kim, S.W., Foissac, S., Lipson, D., Hart, C., Roels, S., Borel, C., Antonarakis, S.E., Monaghan, A.P., *et al.* (2010). New class of gene-termini-associated human RNAs suggests a novel RNA copying mechanism. Nature *466*, 642–646.

Mardis, E.R. (2008). The impact of next-generation sequencing technology on genetics. Trends Genet. *24*, 133–141.

Margulies, M., Egholm, M., Altman, W.E., Attiya, S., Bader, J.S., Bemben, L.A., Berka, J., Braverman, M.S., Chen, Y.J., Chen, Z., *et al.* (2005). Genome sequencing in microfabricated high-density picolitre reactors. Nature *437*, 376–380.

Morozova, O., and Marra, M.A. (2008). Applications of next-generation sequencing technologies in functional genomics. Genomics *92*, 255–264.

Nutiu, R., Friedman, R.C., Luo, S., Khrebtukova, I., Silva, D., Li, R., Zhang, L., Schroth, G.P., and Burge, C.B. (2011). Direct measurement of DNA affinity landscapes on a high-throughput sequencing instrument. Nat. Biotechnol. *29*, 659–664.

van den Oever, J.M., Balkassmi, S., Verweij, E.J., van Iterson, M., Adama van Scheltema, P.N., Oepkes, D., van Lith, J.M., Hoffer, M.J., den Dunnen, J.T., Bakker, E., *et al.* (2012). Single molecule sequencing of free DNA from maternal plasma for noninvasive trisomy 21 detection. Clin. Chem. *58*, 699–706.

Orlando, L., Ginolhac, A., Raghavan, M., Vilstrup, J., Rasmussen, M., Magnussen, K., Steinmann, K.E., Kapranov, P., Thompson, J.F., Zazula, G., *et al.* (2011). True single-molecule DNA sequencing of a pleistocene horse bone. Genome Res. *21*, 1705–1719.

Ozsolak, F. (2012). Third-generation sequencing techniques and applications to drug discovery. Expert Opin. Drug Discov. *7*, 231–243.

Ozsolak, F., Kapranov, P., Foissac, S., Kim, S.W., Fishilevich, E., Monaghan, A.P., John, B., and Milos, P.M. (2010a). Comprehensive polyadenylation site maps in yeast and human reveal pervasive alternative polyadenylation. Cell *143*, 1018–1029.

Ozsolak, F., and Milos, P.M. (2011). RNA sequencing: advances, challenges and opportunities. Nat. Rev. *12*, 87–98.

Ozsolak, F., Platt, A.R., Jones, D.R., Reifenberger, J.G., Sass, L.E., McInerney, P., Thompson, J.F., Bowers, J., Jarosz, M., and Milos, P.M. (2009). Direct RNA sequencing. Nature *461*, 814–818.

Ozsolak, F., Ting, D.T., Wittner, B.S., Brannigan, B.W., Paul, S., Bardeesy, N., Ramaswamy, S., Milos, P.M., and Haber, D.A. (2010b). Amplification-free digital gene expression profiling from minute cell quantities. Nat. Methods *7*, 619–621.

Rumble, S.M., Lacroute, P., Dalca, A.V., Fiume, M., Sidow, A., and Brudno, M. (2009). SHRiMP: accurate mapping of short color-space reads. PLoS computational biology *5*, e1000386.

Sanger, F., and Coulson, A.R. (1975). A rapid method for determining sequences in DNA by primed synthesis with DNA polymerase. J. Mol. Biol. *94*, 441–448.

Tessler, L.A., Reifenberger, J.G., and Mitra, R.D. (2009). Protein quantification in complex mixtures by solid phase single-molecule counting. Anal. Chem. *81*, 7141–7148.

Thompson, J.F., Reifenberger, J.G., Giladi, E., Kerouac, K., Gill, J., Hansen, E., Kahvejian, A., Kapranov, P., Knope,

T., Lipson, D., *et al.* (2012). Single-step capture and sequencing of natural DNA for detection of BRCA1 mutations. Genome Res. *22*, 340–345.

Valouev, A., Ichikawa, J., Tonthat, T., Stuart, J., Ranade, S., Peckham, H., Zeng, K., Malek, J.A., Costa, G., McKernan, K., *et al.* (2008). A high-resolution, nucleosome position map of C. elegans reveals a lack of universal sequence-dictated positioning. Genome Res. *18*, 1051–1063.

Yu, M., Ting, D.T., Stott, S.L., Wittner, B.S., Ozsolak, F., Paul, S., Ciciliano, J.C., Smas, M.E., Winokur, D., Gilman, A.J., *et al.* (2012). RNA sequencing of pancreatic circulating tumour cells implicates WNT signalling in metastasis. Nature *487*, 510–513.

SNP Assessment on Draft Genomes from Next-generation Sequencing Data

3

Michael Piechotta and Christoph Dieterich

Abstract

The number of genome projects has dramatically increased with the advent of high-throughput sequencing technologies. Typically, most assemblies remain in draft status. We present two methods for robust polymorphism detection on draft genomes. The first application, *ACCUSA,* identifies true SNPs on draft genomes by considering the reference base call quality. We exemplify the advantages of ACCUSA by using publicly available sequencing data from yeast strains. Furthermore, we present a *novel* method ACCUSA2 that allows to identify SNPs between sequenced samples and show that *ACCUSA2* outperforms a state of the art SNP caller in an *in silico* benchmark.

Background

Model organisms have been originally selected for controlled experiments in developmental biology and physiology. The success of model organisms is closely linked to the availability of versatile tools for genetic manipulation and the applicability of functional genomics approaches. However, laboratory organisms are often devoid of genetic variability and lack important phenotypic traits. The sampling of model species is strongly biased and several animal phyla are poorly represented in the collection of current model systems.

Several aspects of evolutionary biology and ecology cannot be adequately addressed with the current panel of model organisms. Evidently, non-model organisms may have an interesting biology but have not been the target of systematic molecular analysis due to the absence of tools or prohibitive costs. Ekblom and Gallindo (2011) and Nawy (2012) review recent developments in this area.

Non-model organism have become more accessible for molecular analyses by sequencing. The throughput of next-generation sequencing is rising and the costs per base bisect every 5 months (Stein, 2010). This development facilitates the realization of sequencing projects for non-model organisms even for smaller research groups on tight budgets. However, assembling genomes from shotgun reads is still a daunting task. At the time of writing, software solutions for automated genome assembly deliver draft assemblies of reasonable quality from short read data (Earl *et al.,* 2011; Salzberg *et al.,* 2012).

Nevertheless, genome finishing remains a manual endeavour, which is usually skipped due to time and financial constraints. That is why, most genome assemblies remain in draft status.

However, the draft data could be exploited for the discovery of polymorphisms or sequence variants between individuals, isogenic lines or strains.

Large-scale genomic comparisons of non-model organisms among and within species will soon become reality, and we look forward to contributing to these developments by supporting polymorphism discovery in this very setting.

Single nucleotide polymorphisms (SNPs)

According to Brookes (1999), a single nucleotide polymorphism (SNP) is defined as a single base pair position in genomic DNA, at which different sequence alternatives (alleles) exist in individuals from some population(s). SNPs can be found in the coding region of the genome where they may affect the proper processing or function of gene products, but SNPs can also be found in the non-coding regions of the genome where they may affect regulatory mechanisms. Alternatively, SNPs could be neutral yet are genetically linked to interesting phenotypic changes and thus provide useful markers for genome-wide association studies (GWAS).

Marth *et al.* (1999) proposed to use sequencing data, from shotgun DNA sequencing machines, to mine for SNPs. Their idea was to perform anchored multiple alignments of sequencing reads and to use base quality values to discern true SNPs from sequencing errors. The probability of a SNP is modelled using Bayesian statistics, where the Phred base quality values of a specific column provide the input and a SNP probability is calculated as an output.

The Bayesian model used for SNP screening in the PolyBayes approach (Marth *et al.*, 1999) is a very thorough and consistent way to do SNP screening. However, this approach was developed to work on high-quality EST sequence data and is not directly applicable to deep-sequencing data or genomes in draft status. Whereas the first limitation is mainly due to the very high coverage and the higher error rates in deep-sequencing data compared to Sanger sequencing data, the second issue is mainly due to incorrect consensus base calls within the draft assembly which is later used for read mapping.

SNP calling with one sample on draft genomes with *ACCUSA*

In the following, we will present *ACCUSA* – a simple, yet elegant extension of the PolyBayes formalism towards deep-sequencing on draft genomes (Fröhler *et al.*, 2010). The main idea in *ACCUSA* is to consider read quality information from short read sequencing and quality information from the reference genome at the same time. Fig. 3.1 depicts the typical short read stack aligned to the corresponding reference base, which is then employed to call SNPs by the *ACCUSA* software.

Typically, the quality score (QS) of a base call (BC) is represented as a Phred quality score that encodes the probability

$$P(BC\ is\ wrong)$$

of a wrong base call (Ewing and Green, 1998). This score was originally developed within the Human Genome Project:

$$QS_{Phred} = -10\log_{10} P(BC\ is\ wrong)$$

Internally, *ACCUSA* slides along the reference and calculates a base pileup from reads that cover

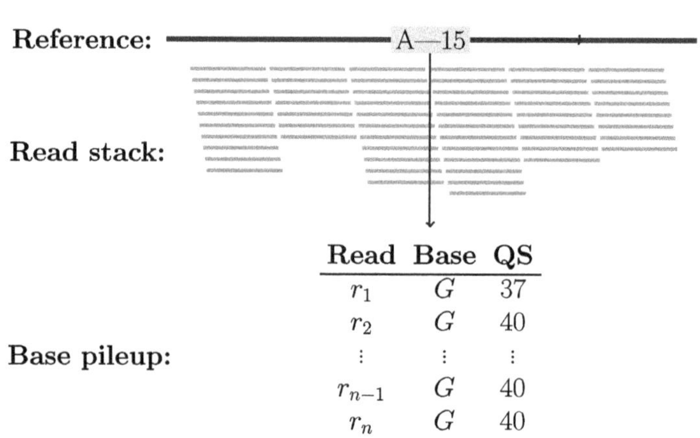

Figure 3.1 Position specific base pileup.

the position under consideration. The relationship between P_{reseq} and P_{all} including the reference quality is shown below:

$$P(\text{reseq}) \left\{ \begin{array}{cc} A & 28 \\ A & 32 \\ A & 34 \\ \vdots & \vdots \\ A & 28 \\ A & 32 \\ C & 15 \\ A & 31 \end{array} \right\} \text{Re-sequencing bases}$$
$$\begin{array}{cc} T & 17 \end{array} \} \text{ Reference base}$$

Finally, the calculated probabilities are filtered according to thresholds defined by the user to give a list of high quality putative SNPs:

$$\text{if } P_{reseq} < P_{all} \wedge P_{reseq} \leq Thr \quad \text{call putative SNP}$$
$$\text{else dismiss SNP}$$

ACCUSA accepts the FASTQ format (http://maq.sourceforge.net/fastq.shtml) to read in the draft assembly sequence and associated quality score information. Short read re-sequencing data must be aligned to the draft genome and must be supplied in the BAM format.

Practical steps

ACCUSA is programmed in JAVA 6 and runs on any platform with JAVA support. The typical *ACCUSA* analysis requires two files as input: a reference sequence in FASTQ file format and short read alignment data in a sorted and indexed BAM file. Generally, all widely used short read mappers provide mapping results in the SAM file format (http://samtools.sourceforge.net/SAM1.pdf).

A unordered SAM file can be converted into an ordered and indexed BAM file with the following *SAMtools* commands (Li *et al.*, 2009):

1 Convert SAM to BAM:
  > *samtools view -bS sample.sam > sample.bam*
2 Sort BAM:
  > *samtools sort sample.bam sample.sorted*
3 Index BAM:
  > *samtools index sample.sorted.bam*

ACCUSA parameters may be adjusted in the *configure.xml* file. The software itself is then executed by typing into a command line interface:

  > java-Xmx2G -jar ACCUSA.jar reads.sorted.bam *min-prob min-depth configure.xml* ref.fastq > result.txt

The *min-prob* parameter is a lower boundary of P_{all}. All positions, which exceed this cutoff are considered in the actual SNP calling procedure. The *min-depth* parameter sets the minimal required short read base coverage for any draft assembly position to be included into the analysis. The *ref.fastq* argument is the corresponding draft genome assembly with quality values. The output of ACCUSA is diverted into *result.txt* and is formatted (see Table 3.1).

Table 3.1 Content of ACCUSA result.txt.

ContigID	position	pSolexa	pAll	avgBpQ	minBpQ	maxBp	Bp of SNP	QS of SNP
Contig0.1	547	7.76E-7	0.99968	45.0	30	77	tt**a**	30 30 **77**
Contig0.2	4430	5.26E-7	0.99999	43.0	30	97	cccc**t**	30 30 30 30 **97**
Contig0.2	7767	6.19E-7	0.99999	32.0	5	97	acaaa**g**	5 5 30 30 30 **97**
Contig89.2	2147	6.19E-7	0.99999	46.0	30	97	tt**c**	30 30 30 **97**
...	...	...	...	...	...	...	...	...

The first column lists the assembly contig. The second column shows the actual nucleotide position (1-based). The pSolexa column is the SNP probability of the short read stack. The pAll column is the SNP call probability in the complete read stack. Reference bases and quality values are always shown at the last position in 'Bp of SNP' and 'QS of SNP' (here emphasized in bold letters).

Availability

ACCUSA is available free of charge to academic users and may be obtained from https://bbc.mdc-berlin.de/software

Example: yeast strain sequencing with draft reference assembly

We used the data of Liti *et al.* (2009) as one way to assess the performance of *ACCUSA*. Liti *et al.* conducted a comprehensive study of genomic variation and evolution by analysing sequences from multiple strains of two Saccharomyces species: *Saccharomyces cerevisiae* and *Saccharomyces paradoxus*. Several strains were sequenced at a depth of 1–3× coverage with conventional Sanger sequencing. Illumina short reads were generated for four *S. cerevisiae* strains and 10 *S. paradoxus* (23–39× coverage). See the Saccharomyces Genome Resequencing Project website for additional details (http://www.sanger.ac.uk/research/projects/genomeinformatics/sgrp.html).

Illumina short reads were mapped on a PCAP assembly of Sanger reads of the Yeast reference strain S288c. All SNP calling coordinates are reported as positions on this draft genome assembly. Liti *et al.* (2009) provided lists of Sanger verified SNPs in chromosomal coordinates based on the finished yeast genome assembly. To compare our SNP predictions with verified SNPs, we had to map coordinates from the draft genome assembly to the finished genome sequence.

We collected a total of 18,956 verified SNPs for the yeast strain W303. From these, 6153 SNPs could be mapped to positions on the assembled draft genome, which are also covered by at least three short reads (covered SNPs). We only consider SNP positions on the draft genome, which are covered by short reads, to compute the precision and recall values of the respective SNP callers. We cannot predict SNPs in the absence of short read alignments or missing genome sequence. Our performance evaluation is shown in Fig. 3.2.

Head-to-head comparisons of sequenced samples with *ACCUSA 2*

In a novel approach we aim to directly identify polymorphic positions between sequenced samples that are mapped to the same reference sequence. This new method exclusively utilizes the base call error probabilities from next-generation sequencing (NGS) platforms to identify and distinguish putative polymorphic positions from sequencing errors.

In *ACCUSA2*, a parallel pileup for each sample is created from sequenced reads (Piechotta and Dieterich, 2013). The parallel pileups are used in a head to head comparison (see Fig. 3.3) to identify polymorphic positions by transforming the base call quality score into probability vectors that are further processed and used by a Dirichlet distribution for statistical testing:

Given the base pileups and the corresponding base quality scores, a probability vector

$$P_{BC} = (p(A), p(C), p(G), P(T))$$

for each base call (BC) is calculated by translating the quality score *QS* to the error probability *P (BC is wrong)* and calculating the probability *P (BC is correct)* that the base call is correct by using the following relations:

$$P(BC\ is\ wrong) = 10^{-\frac{QS}{10}}$$

$$P(BC\ is\ correct) = 1 - P(BC\ is\ wrong)$$

The probability vectors of a parallel pileup are stacked row-wise to form the probability matrix M_k, where $p_{1,C}$ gives the probability that the sequenced base for read R_1 is the base C:

$$M_k = \begin{matrix} Read \\ R_1 \\ R_2 \\ \vdots \\ R_i \\ \vdots \\ R_{n-1} \\ R_n \end{matrix} \begin{pmatrix} A & C & G & T \\ p_{1,A} & p_{1,C} & p_{1,G} & p_{1,T} \\ p_{2,A} & p_{2,C} & p_{2,G} & p_{2,T} \\ \vdots & \vdots & \vdots & \vdots \\ p_{i,A} & p_{i,C} & p_{i,G} & p_{i,T} \\ \vdots & \vdots & \vdots & \vdots \\ p_{n-1,A} & p_{n-1,C} & p_{n-1,G} & p_{n-1,T} \\ p_{n,A} & p_{n,C} & p_{n,G} & p_{n,T} \end{pmatrix}$$

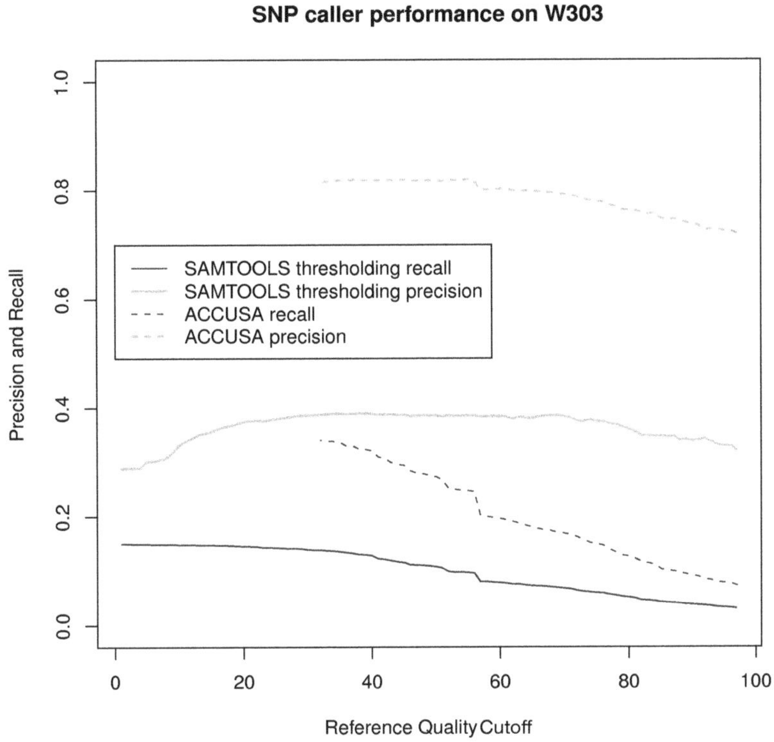

Figure 3.2 Reverse cumulative recall and precision curves for ACCUSA and SAMtools. SAMtools predictions are filtered by a quality clipping step to remove 'error-prone' variant calls. This plot indicates that ACCUSA performs better than a simple quality masking of genomic positions on the set of SAMTOOLS SNP predictions.

We calculate estimates Φ_k of the allele frequency for each pileup by taking the column means of M_k. We model Φ_k with a Dirichlet distribution

$$D(\alpha_k):\alpha_k\left(\alpha_A^k,\alpha_C^k,\alpha_G^k,\alpha_T^k\right)$$

and estimate the parameter vector a_k following a Bayesian strategy.

Polymorphic positions are identified by a likelihood ratio test where

$$L(\alpha_k,\Phi_k):k\in 1,2$$

are the corresponding sample specific likelihood functions and the test statistic λ is defined as follows:

$$\lambda=\frac{L(\alpha_{1,}\Phi_2)\cdot L(\alpha_{2,}\Phi_1)}{L(\alpha_{1,}\Phi_1)\cdot L(\alpha_{2,}\Phi_2)}$$

At the time of writing, we have not released a final version of this methodology, but provide a robust beta version of our software.

Practical steps

A typical *ACCUSA2* analysis run requires two sorted and indexed BAM files as input. The BAM files contain short read alignments of the two sequenced samples (e.g. isolates or variant strains). The draft genome assembly is only employed in the read mapping step to build up the read stacks in the correct positions. *ACCUSA2* produces a file that contains putative polymorphic positions between samples. Depending on the employed mapping tool the following steps are needed to establish the necessary files in order to run *ACCUSA2*:

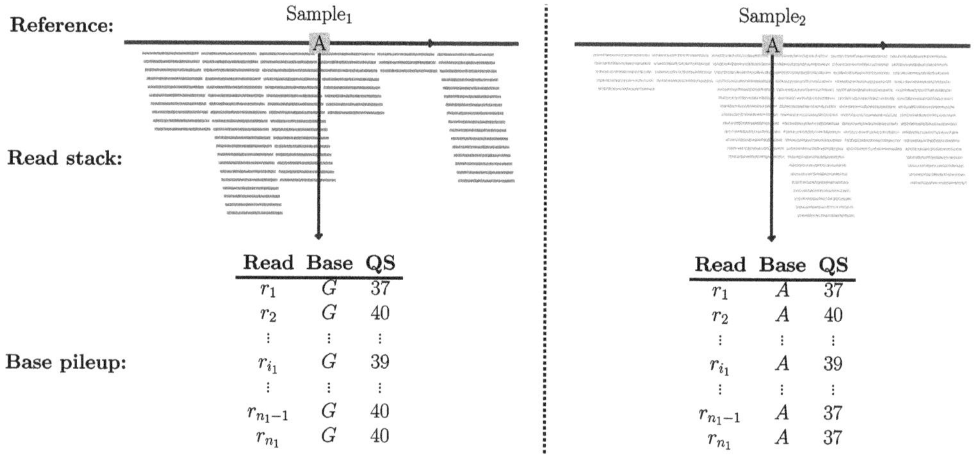

Figure 3.3 Parallel pileups of sequenced samples used for head-to-head comparisons.

1 Convert SAM to BAM
   > samtools view -Sb sampleA.sam > sampleA.bam
   > samtools view -Sb sampleB. sam > sampleB. bam
2 Sort BAM
   > samtools sort sampleA.bam sampleA.sorted
   > samtools sort sampleB.bam sampleB.sorted
3 Index BAM
   > samtools index sampleA.sorted.bam
   > samtools index sampleB.sorted.bam

After pre-processing of SAM files *ACCUSA2* can be started:

```
> java -Xmx4G -jar ACCUSA2.jar -1 sampleA.sorted.bam -2 sampleB.sorted.bam -r output.txt -t 10
```

where *–t 10* retains putative polymorphic sites with a test statistic λ higher than 10. Higher values of λ indicate a higher probability of a polymorphic site. The result file contains the identified polymorphic positions along with base and quality score information for each sample and the test statistic that represents how divergent the sites are.

Example: SNP calling on two synthetic diploid genome

In order to evaluate the performance of *ACCUSA2* on draft genomes we perform an *in silico* benchmark. We create a diploid draft genome assembly from the genome sequence of C. elegans chromosome I (see Fig. 3.4). Synthetic short reads were generated with the *maq* software (Li *et al.*, 2008) to an approximated depth of 10× coverage. These simulated reads are assembled into a synthetic draft genome with the *velvet* software (Zerbino *et al.*, 2008). In the next step, we simulate a mutated sequencing sample with *maq*. We map the mutated read set against the draft assembly that was created from the initial unmutated read set. The two read sets constitute the input for the discovery of polymorphic positions and the implanted mutations are used as a 'gold' standard to assess the performance of SNP callers.

In total, we simulated 30,160 heterozygous SNP positions. Out of these, 27,882 are represented in the draft assembly. We compared the performance of *ACCUSA2* to the *SAMtools/BCFtools* pipeline as outlined in http://samtools.sourceforge.net/mpileup.shtml:

```
> samtools mpileup -uf ref.fa aln1.bam aln2.bam | bcftools view -v -1 1 - > out.vcf
```

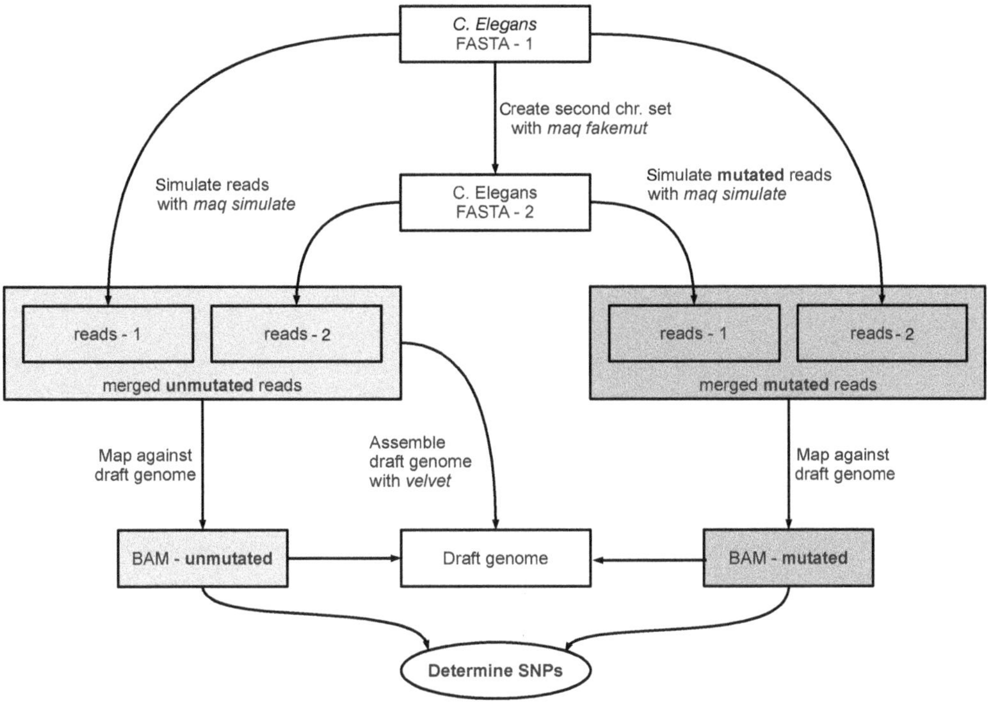

Figure 3.4 Schematic description of the synthetic benchmark.

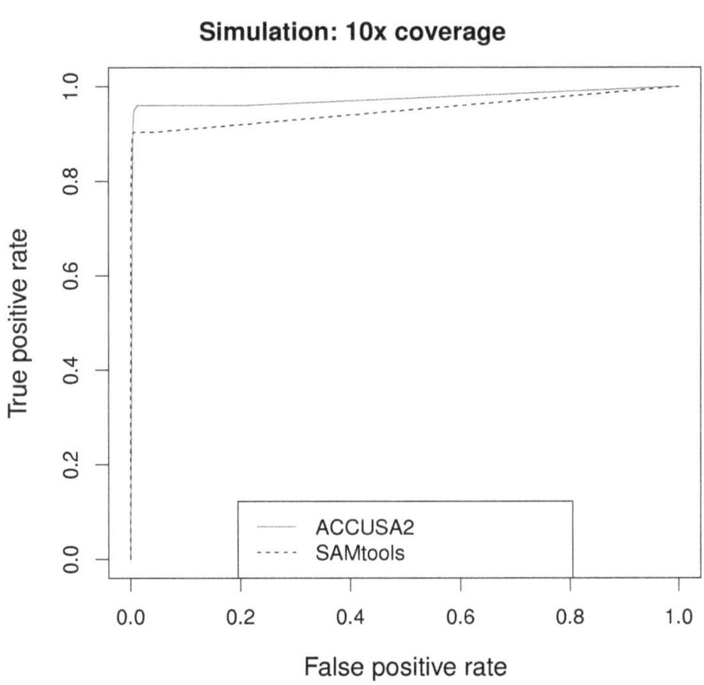

Figure 3.5 Receiver operator curve comparing SAMtools and ACCUSA2.

We used the Receiver operator characteristic curve (ROC) to compare the performance of tested SNP callers on the described benchmark with measures defined as follows:

$$\textit{false positive rate} = \frac{\textit{false positive}}{(\textit{false positive} + \textit{true negative})}$$

$$\textit{true positive rate} = \frac{\textit{true positive}}{(\textit{true positive} + \textit{false negative})}$$

Our performance evaluation is shown as a ROC curve in Fig. 3.5. *ACCUSA2* shows a good performance with an area under curve (AUC) of 0.97 compared to the *SAMtools* with an AUC of 0.94. The AUC indicates how well the SNP caller can distinguish between polymorphic and non-polymorphic sites.

Availability

ACCUSA2 can be obtained upon request from the authors of this chapter (michael.piechotta@mdc-berlin.de or christoph.dieterich@mdc-berlin.de).

Conclusions

We have presented two solutions to SNP calling on draft genomes. The first solution, *ACCUSA* (Fröhler and Dieterich, 2010), was primarily designed to detect haploid or homozygous SNPs in a setting where a short read sequencing sample was aligned to an available draft genome sequence. We have shown a favourable performance on a dataset of resequenced yeast strains. The second solution, *ACCUSA2* (Piechotta and Dietrich, 2013), uses an available draft genome for anchoring short read alignments, but compares directly two short read stacks from two different sequencing samples. We have validated its performance on a synthetic benchmark of heterozygous SNPs. We are certain that both software solutions support the utility of high-throughput sequencing in non-model organisms.

References

Brookes, A.J. (1999). The essence of SNPs. Gene *234*, 177–186.

Earl, D., Bradnam, K., John, J.S., Darling, A., Lin, D., Fass, J., Yu, H.O.K., Buffalo, V., Zerbino, D.R., Diekhans, M., *et al.* (2011). Assemblathon 1: a competitive assessment of de novo short read assembly methods. Genome Res. *21*, 2224–2241.

Ekblom, R., and Galindo, J. (2011). Applications of next-generation sequencing in molecular ecology of non-model organisms. Heredity (Edinb.) *107*, 1–15.

Ewing, B., and Green, P. (1998). Base-calling of automated sequencer traces using phred. II. Error probabilities. Genome Res. *8*, 186–194.

Fröhler, S., and Dieterich, C. (2010). ACCUSA–accurate SNP calling on draft genomes. Bioinformatics *26*, 1364–1365.

Li, H., Ruan, J., and Durbin, R. (2008). Mapping short DNA sequencing reads and calling variants using mapping quality scores. Genome Res. *18*, 1851–1858.

Li, H., Handsaker, B., Wysoker, A., Fennell, T., Ruan, J., Homer, N., Marth, G., Abecasis, G., Durbin, R., and Subgroup, 1000. G.P.D.P. (2009). The Sequence Alignment/Map format and SAMtools. Bioinformatics *25*, 2078–2079.

Marth, G.T., Korf, I., Yandell, M.D., Yeh, R.T., Gu, Z., Zakeri, H., Stitziel, N.O., Hillier, L., Kwok, P.Y., and Gish, W.R. (1999). A general approach to single-nucleotide polymorphism discovery. Nat. Genet. *23*, 452–456.

Nawy, T. (2012). Non-model organisms. Nat. Meth. *9*, 37.

Piechotta, M., and Dieterich, C. (2013). ACCUSA2: multi-purpose SNV calling enhanced by probabilistic integration of quality scores. Bioinformatics *29*(14), 1809–1810.

Salzberg, S.L., Phillippy, A.M., Zimin, A., Puiu, D., Magoc, T., Koren, S., Treangen, T.J., Schatz, M.C., Delcher, A.L., Roberts, M., *et al.* (2012). GAGE: A critical evaluation of genome assemblies and assembly algorithms. Genome Res. *22*, 557–567.

Stein, L.D. (2010). The case for cloud computing in genome informatics. Genome Biol. *11*, 207.

Zerbino, D.R., and Birney, E. (2008). Velvet: algorithms for de novo short read assembly using de Bruijn graphs. Genome Res. *18*, 821–829.

Processing Large-scale Small RNA Datasets *in Silico*

4

Daniel Mapleson,[1] Irina Mohorianu,[1] Helio Pais, Matthew Stocks, Leighton Folkes and Vincent Moulton

Abstract

The latest advances in next-generation sequencing technologies have resulted in a dramatic increase in the total number of sequences that can be produced per experiment as well as a significant decrease in sequencing error and bias. These improvements have driven forward both *in silico* and *in vivo* analyses in small RNA (sRNA) research. Until recently, the majority of existing sRNA computational methods focused on the analysis of a particular class of sRNAs, the micro RNAs. However there are several less well characterized classes of sRNAs present in plants, animals and other organisms that may have important biological function. This has prompted the development of novel data-driven approaches for sRNA analysis that are designed to cope with the increase in both the number of sequences and the diversity of information that is extracted.

This chapter reviews these approaches and consists of three main sections. First, we consider the steps required to produce sRNA libraries. After this, a typical workflow for pre-processing the output from sequencing machines is presented. This includes an outline of the state of the art for adaptor removal, read filtering and selection, read mapping, and various approaches to normalize the read abundances. We then present the main computational techniques for sRNA analysis. More specifically, we discuss qualitative statistics for sample checking, biogenesis driven approaches for identification of known and novel sRNAs, and methods for predicting their function. We also give an overview of how correlation tools, developed to predict the types of interactions between sRNAs and their target genes, can refine information from target prediction tools. The chapter concludes with some remarks on how *in silico* sRNA research might evolve in the near future.

[1] These authors contributed equally to this work.

Introduction

RNA molecules not only facilitate the production of protein in the cell but also control how much protein is produced and when. As such, RNA can be categorized as coding RNA (or messenger RNA – mRNA), which can be translated into a protein, and non-coding RNA, some of which regulates protein production. Many classes of non-coding RNA have been discovered over the past decades. Initially, the classes of longer non-coding RNA were detected, such as transfer RNA (tRNA) (Holley *et al.*, 1965), an adaptor molecule that bridges RNA codons and amino-acids, and ribosomal RNA (rRNA), which plays an essential role in the structure and functioning of the ribosome. A new class of non-coding RNAs called small RNAs (sRNAs) (Zamore *et al.*, 2000) was discovered relatively recently. These short sequences, with lengths between 20 and 30 nucleotides (nt), are present in plants (Waterhouse *et al.*, 1998; Hamilton and Baulcombe, 1999), animals (Lee *et al.*, 1993), viruses (Umbach *et al.*, 2008), single-cell organisms (Drinnenberg *et al.*, 2009) and fungi (Nicolas *et al.*, 2010). They can influence the production of proteins at the transcriptional level, by controlling the rate of mRNA production (Verdel *et al.*, 2004), and at the post-transcriptional level, by tightly regulating the rate of mRNA degradation and the cell's ability to translate mRNA into protein (Wu *et al.*, 2006; Humphreys *et al.*, 2005; Antonio *et al.*, 2006).

Next-generation sequencing (NGS) technologies have increased sequencing speed, reduced sequencing costs, and simplified biological sample preparation workflows (Shendure and Ji, 2008). This has revolutionized the way genome research is conducted, making NGS technologies the tool of choice for genome analysis. NGS technologies are not limited to DNA sequencing; RNA sequencing (RNA-Seq), a process generally involving the sequencing of RNA converted into its complementary DNA equivalent (cDNA) (Morin *et al.*, 2008), is frequently used for discovery and profiling of mRNA and sRNA. NGS devices have revolutionised sRNA research by enabling molecular biologists to cost effectively produce sRNA datasets containing millions of reads. Each dataset represents a snapshot of the sRNA population within the cells of a biological sample at a given moment in time (Magi *et al.*, 2010). As sRNA sequencing becomes faster and cheaper over the coming years, this presents a challenge to bioinformaticians who are tasked with efficiently storing, processing and analysing this data. To meet these demands, existing approaches, mainly developed for microarray analysis (Pepke *et al.*, 2009), are being adapted and new software is being produced.

This chapter discusses some of the biological background to sRNAs, the steps required to produce sRNA datasets, and some of the computational tools for handling and analysing this data (Studholme, 2012).

Library preparation and sequencing

Before a NGS machine can sequence sRNA molecules, a library must be prepared according to the device's protocols. Typically this involves a multi-stage process, starting with the collection of either the total RNA input or the pre-isolated sRNAs from a cell. This library then undergoes an adaptor ligation process, which adds, to each end of each fragment, adaptors which are necessary for use during bridge amplification. In many second-generation devices, the ligated molecules then undergo an amplification process based on real-time polymerase chain reaction (RT-PCR) to boost the signal, although for some new third-generation devices amplification is not required (Schadt *et al.*, 2010). Finally, if the sRNAs were not pre-isolated it is necessary to filter the ligated sequences by length to isolate sRNAs.

Ideally the concentration of the sRNA molecules should be preserved by library preparation protocols. However, this was found to not be the case (Linsen *et al.*, 2009; Willenbrock *et al.*, 2009). Indeed, recent studies have found that both ligation (Tian *et al.*, 2010; Hafner *et al.*, 2011; Jayaprakash, 2011; Sorefan *et al.*, 2012) and PCR (Hansen *et al.*, 2010: 5) steps distort concentrations due to differential reaction efficiencies, dependent on nucleotide sequence. In order to mitigate the ligation bias, subsequent protocols use degenerate nucleotides in one or both of the adaptor sequences, increasing the likelihood of each sRNA being efficiently ligated (Jayaprakash, 2011; Sun *et al.*, 2011; Sorefan *et al.*, 2012).

The most widely used NGS platforms are Roche's 454, Illumina's Genome Analyser (formerly Solexa) and ABI's SOLiD. These platforms have different characteristics in regards to, for example, length of reads (highest in 454), throughput (highest in Illumina), and error rates (lowest in Illumina). For an extensive review on this topic see Chapter 1 (Pais *et al.*, 2011). For sRNA sequencing, high-throughput and low error rate are more important than read length since around 35nt suffices to capture sRNAs, which all platforms can achieve. Consequently, Illumina has become the dominant platform in sRNA sequencing (Kozomara and Griffiths-Jones, 2011; Sorefan *et al.*, 2012).

For these platforms, the nucleotide sequencing itself is registered as fluorescent signals. The raw data consists of series of images containing possibly tens of millions of dots, where the colour of each dot identifies the nucleotide present at that location. By using image recognition software, these images are transformed into files containing the nucleotide sequences along with quality information dependant on the strength and colour of each signal. Some third-generation devices use a different technology to sequence molecules. For example, Oxford nanopore's GridION and

MinION devices, pull DNA strands through a hole where a microchip measures minute changes in the electrical current in the membrane around it as individual bases, or pairs pass through (Xie *et al.*, 2012).

Helper tools

Once sequenced, sRNA datasets are stored as a file containing the sequenced reads. Many of these reads will represent actual sRNAs in a biological sample (along with their ligated adaptors), although they may also consist of degradation products, reads mapping to longer forms of RNA or reads that maybe the result of contamination. Typically, the file representing the sRNA dataset is given in FASTQ format, which describes rules for storing information associated with each sequenced read. The FASTQ format has been loosely defined and subsequently had many variants, including, the original Sanger format, Solexa format, Illumina format and ABI Solid's color-space format. Recently there has been an effort to formalize the FASTQ format definition, at least for sequence space data, in order to assist the portability of the files so that they can be run by different parsers (Cock *et al.*, 2010). Briefly, this definition describes the file as being segmented into groups of four lines that describe each read: read title and description; nucleotide sequence; optional repeat line; read quality information.

Using sRNA data in this format can be cumbersome for downstream experiments, which typically are only concerned with two properties: the nucleotide sequence of the read and the abundance of that sequence (i.e. the number of times that distinct sRNA was sequenced). By compressing data into this format the sRNAs can be processed much more efficiently, in terms of both memory and run-time. Exchanging this compressed form of sRNA data between tools and applications requires another formatting standard. The FASTA format, for example (Pearson and Lipman, 1988), is often used for this purpose. The popularity of the format is no doubt attributable to its simplicity; each sequence is represented using two lines, a descriptor line and a sequence line. However, this simplicity limits the power of the format as there is no standardization for record identifiers; thus, title lines can become very long, making them difficult to parse. Despite these limitations, the FASTA format is still commonly used to represent sRNA datasets in situations where data exchange between tools is required.

Whilst the FASTA and FASTQ formats are relatively compact in that they require few characters to describe their structure, they are loosely structured and do not offer much scope for extending the format to include new properties. In addition, the lack of metadata makes it difficult to re-purpose the data for different applications. This suggests that better defined and flexible formats for data exchange might be required in the future.

The remainder of this section describes some of the more common pre-processing steps that can be applied to raw sequencing files in sRNA bioinformatics workflows.

Adaptor removal

Before sequencing, the set of nucleic acid strands are ligated with adaptors at the 5′ and 3′ ends of the molecule. When NGS machines sequence a complementary DNA (cDNA) (representing the original RNA transcripts) that is shorter than the read length of the sequencing device, part of the adaptors are also sequenced. The devices do not automatically trim these adaptors from the sequenced data. Typical read lengths produced by second generation sequencing devices range from 30 nt to 50 nt[2] whereas sRNAs are normally shorter than 30 nt. The majority of reads therefore contain fragments of the adaptor sequence. The adaptor content in the sequenced data varies from sample to sample. Illumina machines, for example, use a single-ended sequencing process where the primer hybridizes with the part of the sequence that corresponds to the 5′ adaptor. Therefore the first sequencing cycle contains the first nucleotide of the sRNA, resulting in reads starting at the beginning of a sequenced strand of nucleic acid but typically containing some of the 3′ adaptor. This is in contrast to 454 pyrosequencing, where generated reads include both 5′ and 3′ adaptor fragments (see Fig. 4.1). For all devices, the

[2] Although reads may be shorter or longer depending on the number of chosen cycles. The read length does not effect the steps and algorithms that can be applied to reveal features in the data.

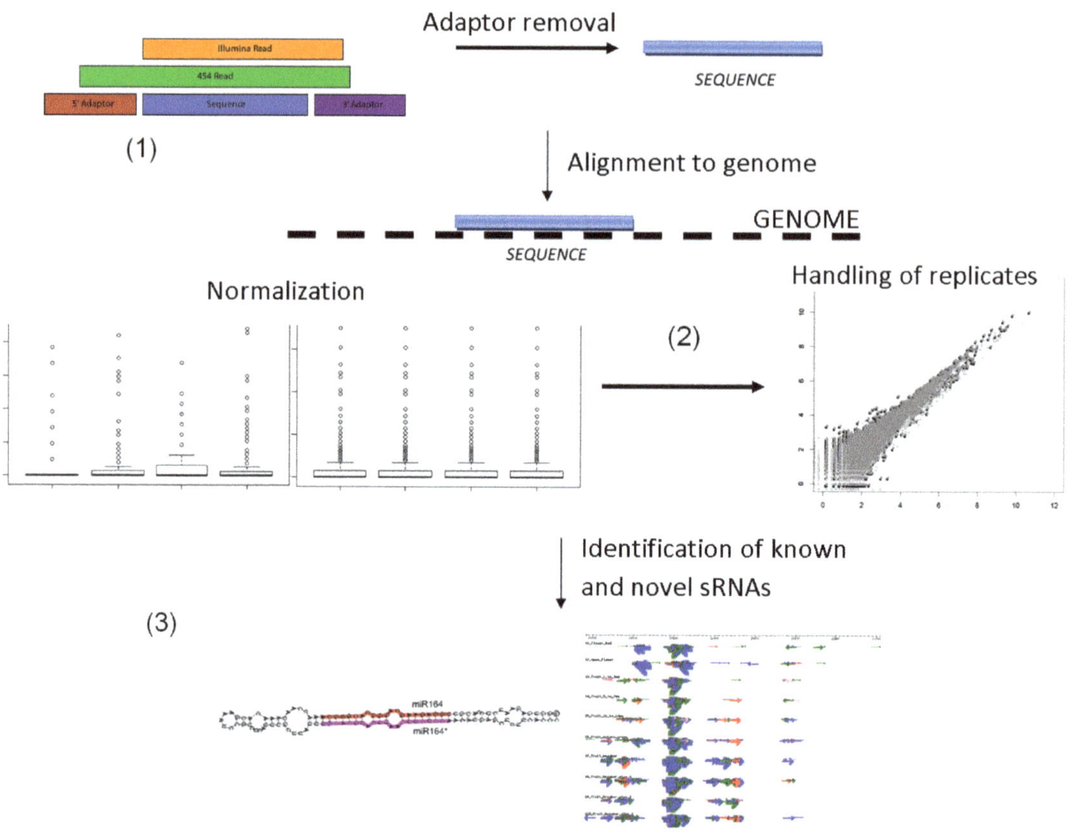

Figure 4.1 Typical sRNA workflow. (1) The reads produced by a next-generation sequencing instrument include adapter fragments that can be removed during the adapter removal step. Also included in the preliminary pre-processing is the matching to a reference genome that can act as a quality control of the sample. (2) Next, the normalization of genome matching reads is performed to ensure that the sRNA abundances are comparable between samples. If replicates are available, an additional quality check is performed using the replicate vs. replicate scatter plot. (3) Subsequent to the pre-processing steps described in 1 and 2 is the identification of known and novel sRNAs. A priori properties such as secondary structure (for miRNAs) or preference for a specific size class (e.g. 24mers as an indication for changing methylation) can be used to partition the reads into different categories. For the presented example, each row contains sRNAs from a sample, each arrow corresponds to an sRNA, with an abundance proportional to the width and the colour illustrates the size class.

trimming of adaptor sequences from sequenced output is a required before further processing of sRNA sequence data.

There is a large selection of open-source adaptor removal tools freely available for use by the community, such as: BTrim (Kong, 2011), the FASTX toolkit Clipper tool (hannonlab.cshl.edu/fastx_toolkit/), Cutadapt (Martin, 2011), DSAP cleanup tool (Huang *et al.*, 2010), the BioStrings BioConductor package (Meur and Gentleman, 2012) and the UEA sRNA Workbench adaptor removal tool (Moxon *et al.*, 2008). These tools have the same underlying principles. The known adaptor sequence(s) and sRNAs are provided to the tool, which then attempts to align the adaptor(s) to each sequence. If a match is found then the adaptor is trimmed. However, the tools vary in terms of the choice of alignment method (global, semi-global) and the type of sequence data supported, such as support for ABI Solid's colorspace data. Also, some adaptor trimmers are more feature rich than others. For example, some trimmers perform general functions such as fuzzy matching, multiple adaptor matching (Martin,

2011), quality trimming, automatic adaptor identification and/or barcode support (Kong, 2011). In addition, some address nuances of particular sequencing machines, such as decreasing signal strength in homopolymer stretches from 454 devices (Nygaard *et al.*, 2009).

Selection/filtering

A common pre-processing step is to isolate sequences with specific properties to focus downstream analysis, improve runtimes and potentially reduce false positive results. For example, if the downstream analysis is focused on micro-RNA (miRNA)s or short interfering RNA (siRNA) s, then discarding sequences with a length less than 20 nt and greater than 24 nt might be beneficial. Another useful property to filter on is the abundance of a sequence within a dataset. The abundance value can be evaluated as a raw value or weighted on the amount of times the sequence will map to the reference genome. In addition, the sequence itself can be normalized through a variety of techniques (see 'Normalization', below) and possibly discarded because of its scaled abundance.

Another technique commonly used to subset an sRNA dataset is, for each distinct sRNA, to look up the sequence to see if it exists in a database. Certain types of known non-coding RNA can be selected or filtered in this way, such as transfer RNA (tRNA), ribosomal RNA (rRNA), small nucleolar RNA (snoRNA) and miRNA. Commonly used databases include Rfam (Gardner *et al.*, 2009; Griffiths-Jones *et al.*, 2003) and miRBase (Griffiths-Jones, 2006, 2010). In addition, users may wish to specify their own list of sRNAs that they either wish to include or exclude in further downstream analysis.

Low-complexity filtering

Formally, the complexity of a sequence can be defined in terms of the length of the sequences shortest description in some fixed universal description language (Kolmogorov, 1965). This is commonly called the Kolmogorov complexity of the sequence. For example, 'GCGCGCGCGC' can be defined 'GC' five times, whereas 'ATGTCGTGCTC' is difficult to represent without using the string itself. The same concept of complexity can be applied to sequences represented in an sRNA dataset.[3] All bioinformatics workflows require low-complexity sequences to be removed. One reason for this is that they often align to transcripts or chromosomes in many locations, thus biasing the downstream analysis and leading to run-time problems.

It is desirable for a low-complexity filter to be both fast and easily understandable. One simple method that satisfies these criteria is to look at the distinct nucleotide count. If the sequence contains a low number of distinct nucleotides (e.g. 1 or 2) then it could be considered to have a low complexity and above this (e.g. 3 or 4) to have high complexity. While this method is generally adequate for sRNAs as it is very unlikely to accidentally remove any sequences that would be useful in downstream analysis, it can retain sequences that still have a low Kolmogorov complexity (Kolmogorov, 1965). For example, if we assume that all sequences with fewer than three distinct nucleotides are considered low-complexity, then 'ATATATATATATATAC' would not be classified as having a low complexity. A number of more sophisticated methods exist that perform better in this regard but this higher accuracy is gained at the expense of runtime and predictability. For example, one approach is to use off-diagonal local-alignment of the sequence against itself (Claverie and States, 1993), another is to look for tandem repeats within the sequence (Benson, 1999), and (S)DUST (Morgulis *et al.*, 2006), uses a heuristic algorithm that employs a scoring function based on counting nucleotide triplets.

Read mapping

A wide variety of techniques have been developed to tackle the sequence alignment problem. Some methods are better suited to certain situations than others. The Needleman–Wunsch algorithm (Needleman and Wunsch, 1970) is perhaps the best known global alignment method, which is useful for aligning sequences of similar length. On the other hand, the Basic Local Alignment Search Tool (BLAST), is the source of the hash table

[3] Albeit with a reduced character set representing the nucleotides in the sequence rather than an alphanumeric character set.

algorithm and is used frequently for identifying species, locating domains, aligning mRNAs to chromosomes and establishing phylogeny (Altschul *et al.*, 1990). BLAST identifies a small seed region with an exact match and then tries to make local alignments around the seed. This is useful for aligning sequences with different lengths, and sequences that share conserved regions (Thompson *et al.*, 1999). However, BLAST is not well suited for aligning short sequences of less than a hundred nucleotides as it uses a heuristic model and is not guaranteed to produce the best possible results.

The production of large sRNA datasets has generated a need for finding perfect or near-perfect matches from short reads to longer reference sequences, typically the genome. There are a plethora of tools for short read mapping that use a range of techniques. Initially, read mapping tools relied on hashing techniques to build an indexing table of oligomers [ELAND, MAQ (Li *et al.*, 2008a), SOAP (Li *et al.*, 2008b)]. A more sophisticated technique uses a pre-processing step based called the Burrows–Wheeler transform (BWT) (Burrows and Wheeler, 1994), which is also known as Block-sorting compression. In this step, reads are transformed into a representation of the original read where none of the actual character values change but are ordered in a way that makes the read more amenable to compression (e.g. run-length encoding). This allows a much more efficient hash table to be built that can result in a fast, memory efficient read mapper that can make good use of modern computer hardware [Bowtie (Langmead *et al.*, 2009), BWA (Li and Durbin, 2009) and SOAP2 (Li *et al.*, 2009b)].

An alternative that is used in the same context is PatMaN (Prufer *et al.*, 2008). The PatMaN algorithm behaves like the Aho–Corasick algorithm (Aho and Corasick, 1975) when searching for perfect matches. In this case the algorithm has a runtime that is linear for the total length of all query sequences. The PatMaN algorithm does allow for mismatches and gaps however the runtime increases exponentially in this case. Compared with the Burrows–Wheeler base read mappers, PatMaN has the advantage that it performs an exhaustive search to identify all occurrences of a large number of short sequences within a genome-sized database, without requiring any indexing operations to be performed up front.

More recently, read mappers have been adapted to make use of alternative hardware, such as GPUs (BarraCUDA (Klus *et al.*, 2012), SOAP3 (Liu *et al.*, 2012)) and FPGAs (VelociMapper, XpressAlign). By exploiting massive parallelism in the hardware, this adaptation can provide significant gains in performance, when compared to conventional CPU processing.

Most of the mapping tools discussed in this section are freely available to download and many are available on multiple platforms.

Normalization

Availability of cheap NGS data has made it feasible to analyse multiple samples, e.g. different organs (Rajagopalan *et al.*, 2006), different mutants (Fahlgren *et al.*, 2009), different responses to stress (McCue *et al.*, 2012) or environmental conditions (Itaya *et al.*, 2008), and different stages of development (Mohorianu *et al.*, 2011). The resulting sets of reads indicate that the sRNAome (i.e. the complete collection of sRNAs in the cell) can significantly change between samples (e.g. in developmental series less than 20% reads are conserved throughout all time points (Mohorianu *et al.*, 2011)). Thus, in order to use sRNA expression series as a basis for exploratory data mining procedures, it is important to identify and minimize the impact of non-biological sources of variation through the normalization of sRNA expression levels (Mccormick *et al.*, 2011) (see Fig. 4.2).

So far, normalization methods proposed for RNA-seq data have been adapted from methods developed for microarrays (Mccormick *et al.*, 2011). Moreover, sRNA studies have mainly focused on the well understood class of miRNAs. A recent study on miRNA-seq data (Garmire and Subramaniam, 2012), classified the commonly used normalization methods into (a) scaling normalizations and (b) distribution-based normalization. The two categories differ in the hypothesis of the source of true biological variation (i.e. variation which can be confirmed in low-throughput experiments such as northern blotting (Mohorianu *et al.*, 2011) or qPCR (Garmire and Subramaniam, 2012)). The first assumes that the

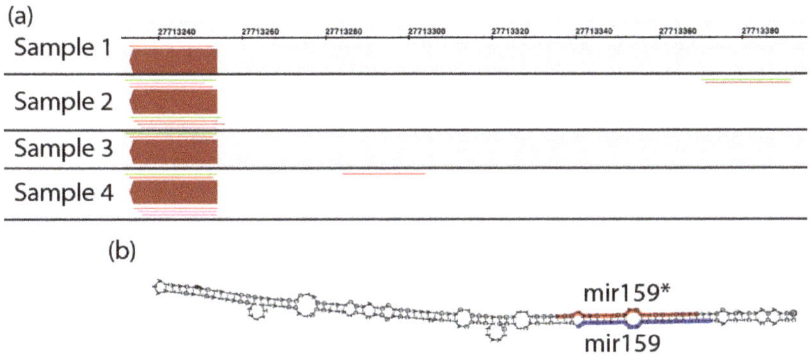

Figure 4.2 miRNA identification using the sRNA Workbench (Stocks *et al.*, 2012) (a) A miRNA locus predicted in four sRNA samples of *A.thaliana* data. The red arrows represent the miRNA. Their thickness is proportional to their abundance and the orientation indicates that the pre-miRNA is transcribed from the negative strand. (b) The secondary structure of the miR159 miRNA pre-cursor obtained using the RNA annotation tool (Stocks *et al.*, 2012). The blue region corresponds to the miRNA and the red region corresponds to the miRNA*.

abundance of the sRNAome is constant across samples and relies on the identification of difference in expression on the proportion of a given sRNA in the total set (or subset) of the reads. The second class of normalization methods presumes that the overall distribution of read abundances is invariant. Whilst bringing the total abundance of reads to a common total (as the scaling normalizations), the abundance proportions of reads are not preserved.

The methods in each category will be briefly described and compared in the rest of this section.

Scaling normalizations

This type of normalization is based on the identification of a global or local scaling factor that is multiplied with all (for global) or a subset of (for local) expression levels of a read within a sample. Included in this category are global normalization (Smyth *et al.*, 2003; Mortazavi *et al.*, 2008), trimmed mean normalization (Robinson and Oshlack, 2010) and Lowess normalization (Smyth *et al.*, 2003).

For global normalization, the scaling factor is defined as the ratio between an arbitrarily chosen total abundance (commonly used value is 1,000,000, hence the synonym for this type of normalization, reads per million (RPM), and the total abundance of reads in the given sample (i.e. the sum of abundances of all reads in a sample). An abundance-based restricted version of the global normalization is the trimmed mean (TMM) normalization proposed by Robinson *et al.* (Robinson and Oshlack, 2010). A scaling factor is computed for all reads with abundances in a predefined abundance interval. This approach is aimed at normalizing the samples using only the reads that are not classified as outliers (high expressions levels) or as noise (low expression levels). Further restrictions based on the intervals of abundances of reads are included in the *locally weighted least squares regression* (Lowess) normalization, which is an example of localized scaling normalization (i.e. local scaling factors are computed for intervals of abundances, selected based on the distribution of abundances within the samples) (Smyth *et al.*, 2003).

Distribution-based normalizations

In contrast to scaling normalizations, which focus on preserving proportions of abundances for reads within the same sample, distribution-based normalizations focus on making the distributions of abundances comparable. Examples include the quantile normalization (Bolstad *et al.*, 2003), variance stabilization methods (Huber *et al.*, 2002) and the invariant set normalization (Pelz *et al.*, 2008; Suo *et al.*, 2010).

Quantile normalization (QN) assigns the same normalized abundance to reads that share the same rank in the different samples forming an experiment. The resulting distributions of abundances for all samples are identical. A localization of identical distribution of abundances forms the underlining presumption for variance stabilization (VSN) and invariant (INV) normalizations. The assumption for VSN arises from microarray studies in which the expression of the majority of genes is unchanged between the different samples forming an experiment. Depending on the choice of reads to be normalized, VSN can also be applied to RNA-seq data. Another variant, a data-driven generalization of VSN is the DDHFm (data-driven Haar–Fisz for microarrays with replication) presented in Motakis *et al.* (2006). It has the advantage of being distribution-free (i.e. does not require a parametric model of intensities) and hence is generally applicable on abundance measurements. INV represents a further generalization of VSN. It is also based on the assumption of constancy of the majority of genes/sRNAs present in the set, but uses the ranked list of abundances instead of the actual measurements (Pelz *et al.*, 2008).

Which normalization method is appropriate?

To date, due to the biases in the different quantification methods, there is no objective reference against which the different normalization methods can be compared (McCormick *et al.*, 2011). Hence, in practice, expression profiles are usually confirmed using qPCR validations (Garmire and Subramaniam, 2012) and northern blot validations (Mohorianu *et al.*, 2011). Of course, the outcome of normalizations can be compared computationally, for example using mean squared error (MSE) (Xiong *et al.*, 2008), where a smaller MSE is preferred to a larger one, since this indicates a better overall outcome. In the Garmire *et al.* study (Garmire and Subramaniam, 2012), using the RPM as reference, the Lowess, QN and VSN consistently produced smaller MSEs. In contrast, the TMM and the INV produced higher MSEs compared even to the un-normalized data (Garmire and Subramaniam, 2012).

These results underline that normalization methods, whilst absolutely necessary, should be used with caution, and resulting expression profiles should be always validated in low throughput experiments. Second, the comparative analysis of normalization methods points out that a search for optimal parameters (or range of values for which the results are stable and 'correct') should be conducted prior to the normalization of reads. Parameters like the intervals of abundance for the Lowess normalization or the limits for the TMM normalization will influence the output and prior to an in depth analysis of the sRNA profiles, the consequences of these variations should be assessed.

Most of these normalization methods are publicly available as R packages: MAD scaling (a variant for the scaling normalization), the QN and the VSN can be found in the *limma* package (limma). Lowess normalization is part of the LPE package (Jain *et al.*, 2003), INV is included in the *affy* package (affy), TMM normalization is included in *edgeR* package (Robinson and Oshlack, 2010), and DDFIFM normalization is included in the homonymous package (Motakis *et al.*, 2006). RPM normalization is included in the UEA sRNA workbench (Stocks *et al.*, 2012).

Analysis tools

Once the sRNA dataset(s) have been pre-processed into a form suitable for analysis, further steps can be conducted to extract salient information from the data. First, the quality of the data can be assessed using some simple statistics, such as the total number sequences, the number of distinct sequences, the size class distribution of the reads and other sequence related properties. Next, to better understand the sRNAs captured in the dataset, two main routes can be followed: (a) the identification of biogenesis-related properties and (b) the identification of function-related properties. The first category consists of the detection of well-defined classes such as miRNAs and trans-acting short-interfering RNA (ta-siRNA)s, and of the identification of reads linked by proximity on the genome and/or similar expression pattern. The latter category includes target prediction and correlation-based studies, aimed at linking sRNAs with the genes that they might regulate. We now discuss these analysis steps in more detail.

sRNA qualitative statistics

The integrity of a sRNA dataset can be assessed using several metrics, such as the total number of sequences, sRNA length distribution, GC content, sequence complexity, and similarity index (Mohorianu, 2012). Other measurements can also reveal particularities of the samples, such as a preference for a specific biogenesis (indicated by the ratio between non-redundant and redundant number of reads, also known as sample complexity (Mohorianu, *et al.*, 2011), or mode of action (indicated by the bias towards a certain size class or known sequence motifs).

The number of reads obtained per sample and subsequently, the number of reads mapping to the reference genome, can be used as a filter for samples that are more likely to contain errors. For example, a very small number of (distinct) reads can underline issues in the sequencing process. In addition, the complexity of a dataset can be computed in terms of the number of redundant (total) versus non-redundant (distinct) reads. This metric indicates a preference for pathways that produce unique mature sequences (such as miRNAs) or for pathways that are characterized by many diverse sequences, linked by other properties, such as size. Also, a high variation in complexity throughout samples may suggest that the samples may not be directly comparable.

The partitioning of the reads into size class (i.e. sequence length) can reveal high-level information concerning the pathways activated in different samples. It can also be used to determine whether or not samples are comparable. For example, a sample with a statistically different size class distribution (under a χ^2 test on the proportion of size classes out of the total) could be either very informative (if the results can be replicated) or indicate technical errors. In either case, it should be treated with caution.

Another quality check, the similarity analysis, involves the identification of overlap between samples i.e. the number of sequences that appear in both compared samples, ignoring the abundance in each sample. A frequently used statistic is the Jaccard similarity index, which can be applied on either all reads in the sample or on the partition per size class. To avoid misleading results coming from low abundance reads, usually a threshold is applied (either an abundance threshold, where only reads with abundances above *t* are considered for the analysis, or a count threshold where only the top *c* most abundant reads are considered. The results of this check can be later supported by a more in-depth differential expression analysis and identification of patterns using unsupervised methods.

A popular tool for analysing FASTQ (and BAM and SAM) files is called FASTQC (Andrews, 2010). This provides useful basic statistics such as total number sequences in the sample, average sequence length, overall GC content percentage, it also provides more detailed statistics such as 'per base sequence content', 'per base GC content', 'per sequence GC content', 'sequence length distribution' and 'sequence duplication levels'. The user may also request overrepresented sequences or highly expressed 5-mers in the sample to be highlights. This tool provides for a good first sanity check of the sRNA sample to see if the content is what might be expected; if it is not then some indication is provided for where further troubleshooting or study is required.

sRNA analysis

Depending on their biogenesis and mode of action, sRNAs in eukaryotes can be categorized as miRNAs, siRNAs and piwi-interacting RNA (piRNA)s. By exploiting known properties of some sRNA classes, it is possible to detect sRNAs from large-scale sRNA datasets that belong to these known classes. Micro-RNAs and ta-siRNAs (a subclass of siRNA) are a good example of this; however, many classes of sRNA do not lend themselves to computation classification in this way. Bacteria, Archaea and Viruses may also contain sRNAs involved in RNA silencing pathways (Murphy *et al.*, 2008) but these mechanisms are different to those mentioned above and are generally less well understood.

Even if the presence of sRNAs (as they are defined in eukaryotes) is debated in bacteria and archaea, a class of sRNA-like elements has been described: the CRISPRs (clustered regularly interspaced short palindromic repeats) (Jansen *et al.*, 2002; Oost and Brouns, 2009). Their main role is to act as genetic memory of unwanted visitors such as plasmids or viruses. This is achieved through storage of a guide sequence (also called seed region or spacer), which is complementary to the invading RNA. The CRISPRs then promote RNA cleavage at a fixed distance of 14 nt from the 3′ end of the guide 'small RNA' (Barrangou *et al.*, 2007; Swarts *et al.*, 2012), thus suppressing the infection. A CRISPR locus is characterized by a succession of 21–47 bp sequences called direct repeats separated by unique sequences of a similar length (spacers). In addition such a locus is generally flanked on one side by a common leader sequence of 200–350 bp. These properties enable the computational classification of CRISPRs by looking for these distinctive patterns in assembled prokaryotic genomes.

RNA viruses may also induce gene silencing. In the event of a virus infection, organisms (both plants (Ratcliff *et al.*, 1997) and animals (Parameswaran *et al.*, 2010)) try to inhibit the viral growth through RNA interference. Making use of their own polymerases, dicers, and argonaute proteins, the invaded organisms promote the production of viral sRNAs (vsRNAs) that can target back the originating transcript (VIGS – virus-induced gene silencing). In addition, it has been shown that as a response to the VIGS, an RNA virus may suppress RNA silencing by depriving eukaryotes of DICER-like enzymes (Burgyn and Havelda, 2011; Takeda *et al.*, 2005). Viral miRNAs have also been found in several DNA viruses (Gottwein *et al.*, 2007), suggesting viral miRNAs can both autoregulate viral mRNAs (Pfeffer *et al.*, 2004) and downregulate cellular mRNAs (Stern-Ginossar *et al.*, 2007).

In case of other eukaryotic sRNA classes, high-throughput sRNA datasets can provide useful information through examination of patterns of sRNAs alignments to genomes or transcripts. sRNAs mapped in this way will often cluster in genomic regions, forming sRNA 'loci', which often display patterns that can help determine their class type. For example, miRNA loci often show distinctive peaks for the mature and star sequences within a locus. The remainder of this section discusses both property-based sRNA classification tools, specifically miRNA classification tools, and locus tools that provide a more general purpose way of analysing sRNAs from NGS datasets.

Micro-RNAs

Non-coding RNA molecules first became associated with gene regulation when sRNA, approximately 22 nt long, were first discovered in *C. elegans* in 1993 (Lee *et al.*, 1993). These sRNAs became known as miRNAs, which are characterized by the hairpin structure formed by the precursor sequence which, in most cases, releases a double stranded RNA (dsRNA) molecule, known as a miRNA duplex. The duplex contains a unique pair of sRNAs, one of which is known as the mature miRNA and the other is known as the miRNA star (miRNA*). The miRNA* typically gets degraded or accumulates at low levels but the mature sequence often goes on to bind, based on complementarity, to a target gene. This either leads to cleavage, resulting in the degradation, or translational repression of the target gene (Wu *et al.*, 2006; Humphreys *et al.*, 2005; Valencia-Sanchez *et al.*, 2006).

Micro RNAs are present in both plants and animals, although they are thought to have evolved through different mechanisms (Axtell *et al.*, 2011), resulting in some significant distinctions, particularly with respect to the targeting and silencing mechanisms (Millar and Waterhouse, 2005). The miRNA sequences themselves are often well conserved across species (Cuperus *et al.*, 2011; Lagos-Quintana *et al.*, 2001; Lau *et al.*, 2001; Pasquinelli *et al.*, 2000), suggesting they appeared early in the evolution of eukaryotic organisms. For miRNAs in both plants and animals, the modes action and biogenesis are better understood than most other types of sRNA and, consequently, the most mature computational sRNA tools today focus on the prediction or analysis of these miRNAs.

There are many examples of computational miRNA tools described in literature and freely available for use by the community. Most use properties of known miRNAs and their precursors to predict novel miRNAs from freshly sequenced sRNA datasets. These approaches use either a rule-based approach, such as miRCat (Stocks *et al.*, 2012) and miRDeep (Friedlander *et al.*, 2008), or a data-driven approach, such as miRPara (Wu *et al.*, 2011). Rule-based approaches all tend to employ a similar strategy: mapping sRNAs to putative precursors and predicting secondary structures using tools such as RNAFold (Hofacker, 2003), then scoring properties according to algorithms based on the standard miRNA biogenesis model. Despite animal and plant miRNAs having significant differences, this has not greatly hindered the development of computational tools for predicting miRNAs. MiRCat (Stocks *et al.*, 2012), for example, offers the ability to switch between rules that favour the prediction of animal miRNAs, to those that might better predict plant miRNAs and vice versa. Data-driven approaches cope with the differences between plant and animal miRNAs by offering two models that each reflect the subtle nuances and characteristics of each type (Wu *et al.*, 2011; Mapleson *et al.*, 2013).

Predicted miRNAs (which ideally should also be experimentally validated) are catalogued in central online repositories. The current primary online repository for miRNAs, is called miRBase (Griffiths-Jones, 2006, 2008, 2010). Fig. 4.2(a) gives a visual representation of the miR159 locus in four different samples of *A.thaliana* sRNA databases predicted using the SiLoCo program (Stocks *et al.*, 2012) at the coordinates given in the miRBase annotation. Fig. 4.2(b) depicts the secondary structure of the miR159 pre-cursor produced using the RNA Annotation tool (Stocks *et al.*, 2012) directly from a miRCat prediction on a set of the same sample data. The miRNA is highlighted in blue and the miRNA* sequence is given in red. Once a miRNA is annotated and stored in miRBase they can then be used by the community to distinguish between the known and potentially novel miRNAs in future experiments. MiRBase stores the mature sequence and the star sequence if present, it also stores the pre-miRNA sequence. Each miRNA often has many variants whether they are found in the same organism or different organisms. These variations reflect minor mutations in the miRNA sequences.

Other sRNAs

Identifying other types of sRNA (i.e. types of sRNA that are not miRNAs) is computationally difficult in most cases. This is because all short-interfering RNAs come from regions (sRNA loci) which produce a larger set of siRNAs based on a dsRNA created from a single-stranded RNA using a RNA dependent polymerase (ta-siRNA (Vazquez *et al.*, 2004; Peragine *et al.*, 2004), hetero-chromatic RNAs (hcRNAs) (Mosher *et al.*, 2008)) or by intermolecular base pairing (nat-siRNAs) (Borsani *et al.*, 2005)). Ta-siRNA prediction, however, is computationally plausible since it is based on the identification of a miRNA initiator target site and characteristic phasing pattern of the sRNA alignments.

In contrast to this, piRNAs, initially found in *Drosophila* germ lines, derived from single-stranded precursors which are currently not well understood (Malone *et al.*, 2009), are difficult to identify due to a lack of conserved secondary structure motifs and sequence homology. However, some conservation can be found at the end of the piRNA itself and recently a computational tool was built to exploit this fact (Zhang *et al.*, 2011). Their function, similar to that of hcRNAs, may be the silencing, at an epigenetic level, of retroviral elements that propagate by infecting neighbouring germ cells. Also, piRNAs exhibit a high complexity (i.e. large number of distinct sequences and variable length) (Seto *et al.*, 2007; Siomi *et al.*, 2011). For this reason, prediction of piRNAs has met with limited success.

Locus tools

Reads forming an RNA-seq sample have two a priori characteristics: their sequence and their abundance. However, neither of these two properties can be used to assign the read to a functional category (Schwach *et al.*, 2009). One way to achieve this goal involves matching the reads to a reference genome as a first step. Besides acting as a filter for potential contaminations (or as an overall quality check of the sample, since a large proportion of the reads should map to the reference genome of the organism to be studied), the genome mapping of reads may provide a clue to their function either through annotation or through structural characteristics. For example, the focus on miRNAs detection was in part driven by the relatively simple structure of their underlining locus (see Fig. 4.2).

As mentioned above, numerous prediction tools, based on classifiers, have been developed to identify miRNA hairpin structures, but also miRNA loci rule-based identifiers have been proposed (e.g. mapMi; Guerra-Assuncao and

Enright, 2010). Another plant specific class of sRNAs, the ta-siRNAs encouraged the development of computational tools for the prediction of the underlining transcript. Algorithms by Chen *et al.* (2007) and Axtell (2010) propose candidates based on the existence of a phasing pattern that is statistically different from random. Additional evidence can be gathered by looking for the target site of the initiator miRNA.

However, in plant datasets, the miRNAs and ta-siRNAs typically form 30% of redundant reads, leaving the characterization of the remaining 70% (Rajagopalan *et al.*, 2006; Mohorianu *et al.*, 2011) as an open problem in computational biology. Currently, three methods have been proposed. Chronologically, the first method is rule-based (Molnar *et al.*, 2007). Briefly, all reads within a predefined distance are grouped in one biological unit. This method started to become obsolete with the increase of sequencing depth and the increase in the number of samples. As more reads were produced per sample, the noise, defined as spurious hits, or fragments that could correspond to random degradation products, favoured the joining of adjacent loci, masking the primary transcripts and thus hindering the identification of specific characteristic properties of loci. The second method, NiBLS (MacLean *et al.*, 2010), was also developed for one sample and defined a locus using a graph based approach. The genome matching sRNAs located within a predefined distance threshold form the vertices of a graph and edges link vertices that are closer than a user-defined distance threshold. Loci correspond to clusters in this graph that have the clustering coefficient above a used defined threshold.

To cope with the increase in the number of available samples these two methods have been adapted to multiple sample experiments by merging the reads in all samples. This results in an increase in the length of loci for the rule-based approach and an over-fragmentation for NiBLS. A third method, SegmentSeq (Hardcastle *et al.*, 2012), splits the genome into non-overlapping fragments which are subsequently classified into loci and nulls using a Bayesian approach based on the abundance patterns of the constituent sRNAs. This approach is capable of handling multiple samples and experiments consisting of replicates. However, it requires a lot of processing time, which increases with the number of samples, sequencing depth and length of the genome.

Following the results on large-scale sequencing experiments (Mohorianu *et al.*, 2011), methods using the patterns of reads are being developed. Using the variation of expression as criteria to define boundaries of loci, the aim is to overcome the effect of low abundance, spurious reads and to predict regions with consistent low level properties, such as size class (Mohorianu *et al.*, 2013).

SegmentSeq is available as R package (Hardcastle, 2012), NiBLS is available as a Perl script (MacLean *et al.*, 2010), and the rule-based and the CoLIde approaches are available as tools in the UEA Workbench (Stocks *et al.*, 2012).

sRNA target prediction

The identification of genes that are targeted by sRNAs plays an important part in understanding sRNA function. Computational methods exist that can carry out *in silico* prediction of sRNA targets and can be loosely placed into three categories: target prediction in plant systems, target prediction in animal systems and degradome assisted target prediction. In plants, sRNAs such as miRNAs tend to bind to mRNAs with near perfect complementarity, which enables tools to utilize the observations made about the binding properties of sRNA/target pairs to identify potential interactions. In plants the interference machinery often results in mRNAs being cleaved, resulting in the degradation of the mRNA. Fragments of the degraded mRNA can be isolated and sequenced on a large scale and the resulting library has often been called the 'degradome'. Tools that utilize a degradome can use the cleavage products as evidence of an interaction when predicting sRNA targets. However, in animals, sRNAs tend to repress rather than cleave targets and it is common for sRNA/mRNA interactions to have many more mismatches than in plant systems. This makes target prediction *in-silico* for animals difficult. We now describe each of these three categories in more detail.

Plant target prediction

In plants, it is possible to predict sRNA targets using our understanding of the binding properties of sRNAs, such as miRNAs, and their corresponding target genes. There is generally a high degree of complementarity between the sRNA and its target, with few mismatches, adjacent mismatches or gaps within the duplex (Schwab *et al.*, 2005). Prediction tools attempt to make a sequence alignment between a sRNA and its potential target genes with consideration of features within the alignment. These features may include the number and position of mismatches within an sRNA/mRNA duplex. Features may also include the adjacency of mismatches, G-U base pairs and gaps or bulges within a duplex. In most target prediction software, the extent to which these features are considered is user configurable, with the user being able to set various thresholds, cut-offs and weightings to make the target search conditions less or more stringent.

Some popular plant target prediction tools include Target-align (Xie and Zhang, 2010), sRNA Target (Moxon *et al.*, 2008), TAPIR (Bonnett *et al.*, 2010) and psRNATarget (Dai and Zhao, 2011). Each tool provides specific features such as high-throughput capability, flexibility of parameter setting or target accessibility evaluation. Furthermore, they provide differing levels of accuracy. As a possible alternative to using a specific prediction software, it is possible to use a database such as StarBase (Yang *et al.*, 2010) to find intersections among targets that have been predicted by multiple programs. However, this brings the disadvantage to the user of being unable to provide their own input sequences. It seems clear that a larger integrated approach is needed that spans the existing target prediction landscape, whilst being flexible enough to incorporate new computational methods as they become available.

Degradome-assisted prediction

As already discussed, in plants, sRNAs, in particular miRNAs, have been shown to bind with near-perfect complementarity to their mRNA targets, generally leading to cleavage of the mRNA. The result of this cleavage is the degradation of the mRNA. However, fragments of the degraded mRNA can remain stable within the cell. The stable degraded mRNA fragments exhibit features such as an uncapped 5′ end with a poly(A) tail at the 3′ end. A biological protocol called parallel analysis of RNA ends (PARE) can be used to sequence mRNA cleavage products on a large-scale (German *et al.*, 2008). This 'snapshot' of cleavage products is known as the 'degradome'.

Computational techniques exist which use the degradome to assist in identifying sRNA targets and provide support for the sRNA–target interaction. The degradome is used in addition to measuring sequence complementarity within a duplex to make target predictions. The first tool to make use of the degradome for this purpose was CleaveLand (Addo-Quaye *et al.*, 2009a), which has successfully identified miRNA targets in a variety of organisms (Addo-Quaye *et al.*, 2008, 2009b; Li *et al.*, 2010; Pantaleo *et al.*, 2010). Other methods that can use the degradome in sRNA-target prediction include SeqTar (Zheng *et al.*, 2012) and SoMART (Li *et al.*, 2012). SeqTar is a method that allows a less stringent alignment criteria between sRNAs and their potential targets than that used in CleaveLand. The method introduces two statistics, one to measure the sRNA/mRNA alignment and another to quantify the abundance of degradome reads at the centre of the sRNA/mRNA duplex. At the time of writing the method is not publicly available in a downloadable package. SoMART (Li *et al.*, 2012) is a web-based collection of tools. In particular, the web-tools called 'Slicer detector' and 'dRNA mapper' can be used in conjunction to identify sRNA–target interactions and evidence those interactions through the degradome.

The algorithms employed by both the CleaveLand pipeline and the SeqTar method are not suited to processing the millions of sRNAs obtained from a high-throughput sequencing experiment within a reasonable time-scale[4]. By the nature of the SoMART web-tools, any large scale data analysis would require additional software to combine the output of the Slicer detector and dRNA mapper tools. Recently, a tool called PAREsnip (Folkes *et al.*, 2012) has become available which can be used for high-throughput sRNA target prediction and uses the degradome to support those predictions. Its algorithms exploit the current biological understanding of how cleavage occurs in nature to quickly discard possibilities

[4] At least without setting up scripts to farm out many small jobs across multiple machines in parallel.

that are unlikely to be genuine sRNA/target interactions. The efficiency of this approach allows users to search for potential targets of all sRNAs obtained from a NGS experiment and presents the possibility of generating complete networks of sRNA/target interactions. PAREsnip also makes efficient use of modern hardware, runs on all popular modern platforms and is user-friendly. It is part of the UEA sRNA Workbench (Stocks *et al.*, 2012) and is supported by other tools within the Workbench that can assist with the visualization of results.

Animal target prediction

Predicting animal sRNA targets *in silico* is difficult due to the less specific and efficient mRNA binding that takes place in comparison to plants. In animals, there are broadly two types of target sites: the first one, called a seed site, contains a region of 6–8 nt perfectly complementary to the first 8 nt on the 5′ end of the sRNA; the second one, called a 3′ compensatory site, contains weaker complementarity across the entire sRNA. Both these types of targets mediate mRNA degradation and translation repression, rather than cleavage. The first computational target prediction methods used sequence complementarity to sRNA, conservation and thermodynamic stability of target sites to identify putative sRNA targets (reviewed in (Sethupathy *et al.*, 2006; Rajewsky, 2006)). The most widely used of these methods are miRanda (Enright *et al.*, 2003), PicTar (Krek *et al.*, 2005) and TargetScan (Lewis *et al.*, 2005). Although these methods were valuable in assisting the validation of many targets, they tend to produce a very large number of false positives. In order to increase the specificity of the target predictions, subsequent approaches have included high-throughput mRNA (microarrays or RNA-seq) and protein data (SILAC), combined with manipulation of sRNA activity level or RISC immuno-precipitation (reviewed in Thomas *et al.*, 2010). Recently, a new method called HITS-CLIP (Darnell *et al.*, 2010) was proposed.

Correlation tools

Another layer of information that can reveal the role of molecules in the cell makes use of variation in expression levels throughout several conditions. Starting with the comparison of mRNA and protein level in 1999 in the Gygi *et al.* (1999) study, and followed by several other studies suxh as that of Greenbaum *et al.* (2003) and Yu *et al.* (2007), the idea was to produce a 'computational validation' of interactions. The mode of action of miRNAs, as negative regulators of mRNA (i.e. through cleavage or translational repression) led to the introduction of comparative analysis of expression levels of sRNAs and genes.

In numerous studies (particularly for animal samples) improvements in target prediction were sought by incorporating the correlation between the miRNA and mRNA expression profiles to discriminate between true and false positives (Huang *et al.*, 2007). However, it has been noticed that only some proportion of the pairs [generally estimated to be around 50% (Nielsen *et al.*, 2009; Wang and Li, 2009; Nunez-Iglesias *et al.*, 2010)] show a statistically significant negative correlation. In addition, examples of miRNAs that are positively correlated with their target genes have only been found in plants so far (Kawashima *et al.*, 2009). Moreover, in plants, it has been shown that other classes of sRNAs (like hcRNAs) can exhibit both positive and negative correlation with the genes that they regulate (Mosher *et al.*, 2008).

These observations raise a question concerning the modes of action of sRNAs. Several modes of action of miRNAs have been reviewed in 2009 by Voinnet (2009) but currently it is not known whether these are exceptions or the rule. A genome wide expression analysis was conducted to make progress in answering this question. This was facilitated by the development of a tool, FiRePat, finding regulatory patterns. Using FiRePat, which uses as similarity measure the Pearson Correlation Coefficient, it was found that the positively and negatively correlated pairs were equally frequent in both plants (*Arabidopsis thaliana*) and human (*Homo sapiens*) samples. In addition, for plant miRNAs, it was discovered that variants of the same miRNA (sequences that belong to the same miRNA family) can be both positively and negatively correlated with genes belonging to the same family and sharing the miRNA target site (Mohorianu *et al.*, 2012).

This finding suggested that (a) the a priori conditions required for traditional correlations

(Pearson, Spearman, and Kendall correlations) were not met in several cases and (b) that the correlation might not be visible throughout the entire set of samples within an experiment. Based on these remarks, a simplified correlation was recently proposed (Lopez-Gomollon *et al.*, 2012), that took into account only the direction of variation and not its amplitude. More specifically, Pearson, Spearman and Kendall correlations were applied on simplified series of small integers, created by assigning +1 when the expression level increases, −1 when it decreases and +0 when the expression levels of two adjacent points do not support differential expression. This simplification produced results that confirmed the previous conclusion that positive and negative correlations are equally frequent, and also allowed the refinement of the pairs categorized as not correlated into not correlated and showing mixed correlation (i.e. containing a window of positive and a window of negative correlations).

It should be noted that correlation studies can be used beyond the scope of improving target prediction tools. In particular, these studies have the potential to reveal indirect targets or short-list members of the same regulatory pathway. Also, they can help in assigning a role to other sRNAs, like hcRNAs. Recent studies, such as that by Baev *et al.* (2010), and the annotation of the tomato genome (Tomato, 2012) revealed that a large proportion of genes are regulated at promoter level, by hcRNAs.

Visualization tools

The combination of NGS data and more efficient analysis tools, is producing vast quantities of results that are not always easy to filter or interpret. Visualization tools offer a way to explore large amounts of data quickly in order to gain new insights. In some cases, they also provide an easier interface that enables users to understand the data, and its related concepts, faster.

At the time of writing, several eukaryotic organisms have had their genomes fully assembled. A decade ago genome sequencing and assembly was a very time-consuming and expensive business as was well documented by the Human Genome Project (Human, 2001). Today, NGS technologies and powerful open-source tools have enabled smaller groups to sequence and assemble complete genomes with a relatively modest budget (Heliconius, 2012; Tomato, 2012). Third-generation sequencing machines promise to make this task easier and hence reference genomes for more species are likely to become available at an accelerated rate in the future. However, a reference genome on its own is of little use. To properly understand them, biologically functional regions must be annotated, such as protein coding genes, proteins, non-coding genes, and regulation sites, as well as sRNA loci.

These annotations, often stored in standardized file formats such as the General Feature Format (GFF), can be visualized using genome browsers. One of the most powerful genome browsers available at the moment is Ensembl (Hubbard *et al.*, 2002; Flicek *et al.*, 2012). This tool is presented as a website, which can also be downloaded, customized and hosted on a local server. It supports a comprehensive range of fully assembled genomes along with extensive annotations and the ability to do comparative genomics. However, because of its web-architecture, the client interface (accessed through a web-browser) does not offer the same user experience as is possible from a desktop application. Tools such as VisSR (Stocks *et al.*, 2012) and Savant (Fiume *et al.*, 2010), offer an interface where the user is able to rapidly move and zoom around each chromosome. In addition, both browsers are designed to enable the dynamic visualization of NGS sequencing data. VisSR in particular is suited to visualizing sRNA data in a genomic context, enabling the user to load a sRNAome and display each read aligned to the genome, making it easier to detect patterns, indicative of some biological function.

Until recently, the sRNA datasets have been visualized by assigning a graphic element to each sRNA e.g. a sRNA is represented by an arrow, for which the colour indicates the length of the sRNA and the thickness indicates the abundance (absolute or relative to a fixed element), as shown in Fig. 4.3. These representations allow the user to identify patterns (corresponding to e.g. sRNA loci) by eye. However, with the recent rapid increase in the number of reads per sample, it has become unfeasible to quickly draw a conclusion

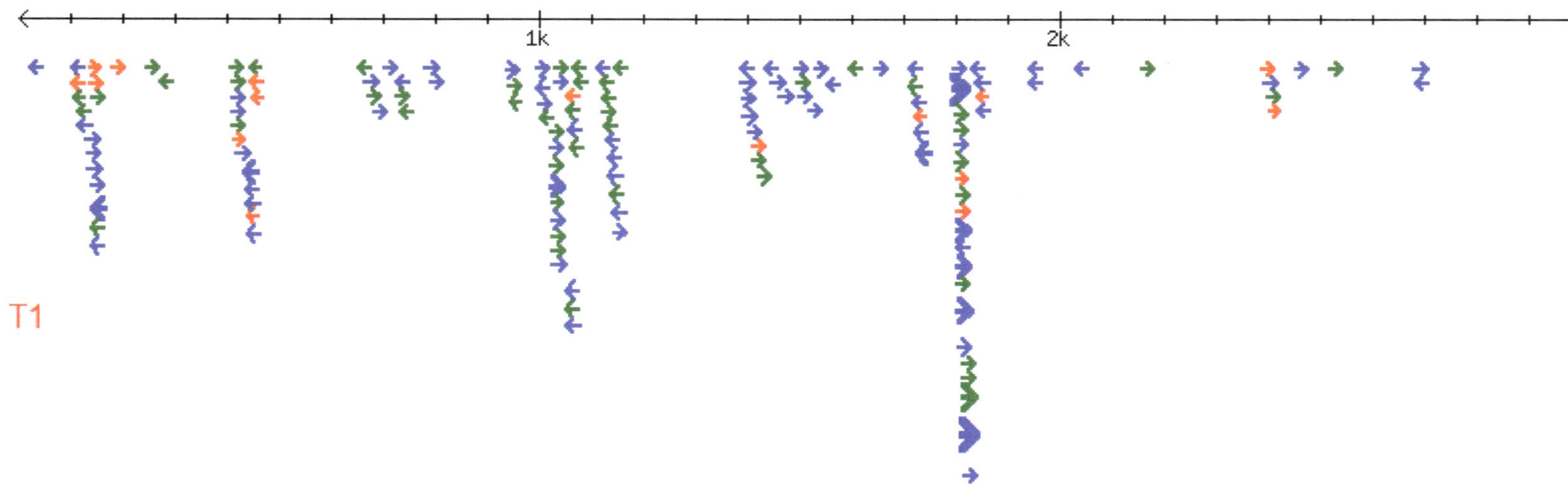

Figure 4.3 Genome browser figure used for a relatively small number of reads, each arrow represents a read with the characteristics described in the caption for Figure 4.1.

just by looking at the distribution of reads. To help the generation of hypotheses, a summarizing representation has been proposed (Mohorianu *et al.*, 2012; Tomato, 2012) that makes use of properties like abundance and size class distribution. Briefly, this involves the genome or transcript of interest being split into windows and for each window a rectangle is drawn. The height of a rectangle is proportional to the summed abundances of the sRNAs incident with the window, i.e. sRNA either completely inside the window or sRNAs for which more than 50% of the sequence is inside the window. For each window, a histogram for the size class distribution of the constituent sRNAs is presented. This is shown in Fig. 4.4.

This initial step towards a condensed presentation of reads changed the focus from the individual reads to the properties that define them. Similar to the transition from the study of miRNAs to the study of general loci, it is no longer crucial to be aware of the precise location of a read, as long as a detailed enough report on the properties of the reads is presented.

We conclude this section by noting that visualization of sRNAs and associated RNA molecules is not limited to browsing NGS datasets. Other examples of visualization include: ta-siRNA phasing analysis (Molnar *et al.*, 2007) and RNA folding and annotation (Stocks *et al.*, 2012). The latter tools serves as a wrapper for the folding and structure plot tools created as part of the Vienna toolkit (Hofacker, 2003), one of several software packages dedicated to RNA sequence analysis.

Discussion

NGS technologies have revolutionized the way sRNA research is conducted. It is now routine to produce sRNA datasets that represent a snapshot of the sRNA content within a biological sample at any moment in time. This has moved many tasks that were previously performed in the wet lab into the bioinformatician's domain and opened up the possibility of conducting more detailed analyses than were previously possible. For example, by taking several snapshots of the sRNA content across different tissues and time points, it is possible to get a much better understanding of the gene regulation processes that take place within an organism. In addition, by analysing the differences in sRNAs, and in particular miRNAs, which are good phylogenetic markers due to their low rate of evolution across organisms, it is possible to improve our understanding of evolutionary biology.

Whilst second-generation sequencing technologies have revolutionized sRNA research, third-generation sequencing machines promise to further increase throughput, decrease the time to result and to increase read lengths. This should also make genome assembly much cheaper, quicker and less complex, which should greatly benefit sRNA research since a lack of reference genomes hinder many current sRNA tools.

Platforms that sequence single molecules, a hallmark of several third-generation devices, should require less bias-introducing library preparation steps, such as PCR, and lead to more accurate estimates of sRNA concentrations. Sequencing bias is however an artefact of Illumina sequencing, and Illumina looks likely to remain the dominant platform in sRNA sequencing for the short to middle term (Sorefan *et al.*, 2012). This bias is a non-biological source of variation, which presents a particular challenge for normalizing sequence counts across samples, prompting the development of approaches that better handle the particularities of count data, rather than simply adapting techniques created for microarrays. However, given the lack of a golden standard, new adaptive approaches might model better each particular dataset. For example, semi-supervised learning could improve normalization steps by using low throughput validations (such as northern blots or qPCR) as controls. This could, for example, be used to improve normalizations methods in the scaling category since scaling factors could be cross-validated using the quantification of the controls. This 'educated normalization' would assure both local consistency (i.e. local normalization or a priori chosen sRNAs) and global consistency (by adjusting the normalization factors to fit all available validations).

The steady increase in the quality and quantity of sRNA sequence data, driven by NGS technologies, represents a challenge for the algorithms developed to analyse this data (Mohorianu

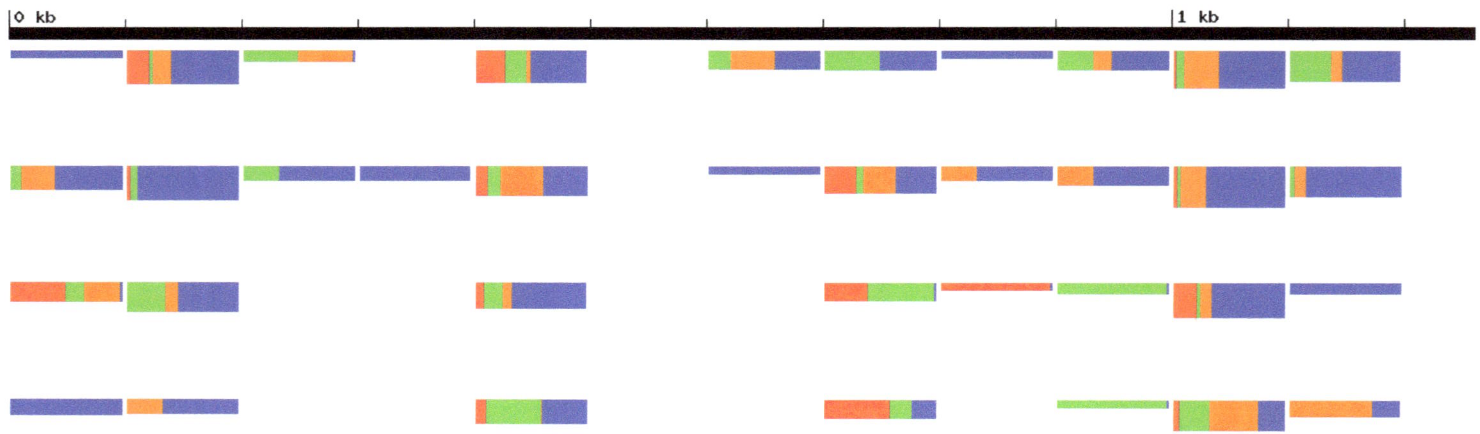

Figure 4.4 Summarizing genome browser figure applicable to a larger number of reads per sample. The genome is partitioned into windows of fixed length (in this example, W=100) and for each window containing incident sRNAs a rectangle is drawn. The height of the rectangle is proportional to the combined abundance of incident reads. The colour histogram for each rectangle reflects the size class distribution. A read is incident to a window if more than 50% of its sequence is contained within the window.

and Moulton, 2010). Faster and more efficient versions are now required to mine the vast datasets that can be readily acquired. For example, miRCat, originally created as a set of web-based Perl scripts (Moxon *et al.*, 2008) has now been updated to take advantage of multi-core processors to dramatically improve run-time over the original implementation of the algorithm (Stocks *et al.*, 2012). Furthermore, new tools should be designed to take advantage of multi-threading technology whenever possible from the outset, as in the case of PAREsnip (Folkes *et al.*, 2012) for example. In parallel, with optimizing existing tools and creating new ones, it is beneficial to incorporate them into easy-to-use pipelines. This, in turn, enables researches to chain multiple tools together, to both better understand their datasets and more quickly convert the data into the format they need for downstream experiments. There have been steps made in this direction, for example a combination of the Galaxy (Goecks *et al.*, 2010) framework and the Taverna (Hull *et al.*, 2006) workflow management system provides the user the opportunity to chain a wide variety of bioinformatic tools (not just sRNA tools) together using a consistent visual interface. However, the interface is purposefully generic and consequently requires the user to have a thorough understanding of the tools to make good use of this solution. Further progress in refining file formats and developing common standards could ameliorate this problem in the future. The SAM/BAM formats (Li *et al.*, 2009a), which are used for exchanging alignment data, are a good example of progress in this direction.

Currently, sRNA class prediction tools are mainly limited to identifying miRNAs and TAS loci in a genome as other classes of sRNA do not easily lend themselves to computational classification. While it is no doubt possible to improve upon the current range of tools in this area, it remains to be seen whether new tools can be (or need to be) developed to detect other classes of sRNAs. To some extent this depends on whether new types of sRNA are detected and proved to interact with target transcripts. New high-throughput target prediction tools (Folkes *et al.*, 2012) and whole sRNAome analysis and correlation tools (Mohorianu *et al.*, 2012) might be capable of this.

Correlation tools involve partitioning a search space into groups that share properties currently relies on choosing a good distance or a good measure of correlation. Thes types of studies, which are currently still in early stages of development, have been applied to specific sequences, such as miRNAs. This approach revealed fragments of regulatory pathways, which were later combined with other un-annotated sequences using clustering methods (Mohorianu *et al.*, 2012). A first step was made with FiRePat, which enables the user to identify groups of sRNAs tightly correlated at expression profile level, which are also correlated to mRNAs. The next step would be the use of annotations (such as GO terms for mRNA) to identify clusters enriched in particular functions. Subsequently, an in depth analysis of the sequence-based interactions may provide further information for establishing the direction of interactions, such as determining the cause and the effect.

Regardless of whether new tools need to be developed or not, it is clear that the task of identifying primary transcripts is not likely to be solved with only one dataset. A reliable map of sRNA loci for an organism will almost certainly be the result of corroborating information from multiple datasets describing different aspects. For example, the concurrent use of developmental series, organ series, mutant series, might provide an insight on the different regulatory layers that involve sRNAs. In addition, existing tools are currently predicting new sRNAs and their targets as more sRNA datasets are produced. This in turn will lead to the number of experimentally validated sRNA targets increasing. It is therefore becoming feasible to use machine learning methods to characterize target sites classes and hopefully more accurately define the sets of genes targeted by each sRNA (Sturm *et al.*, 2010; Reczko *et al.*, 2011; Heikkinen *et al.*, 2011). An alternative approach is to combine the many existing prediction methods to obtain more robust sets of targets (Herrera *et al.*, 2011).

This is an exciting time in sRNA research. As more sRNA samples are sequenced and as hardware and software are refined, our understanding of the mechanisms underlying gene regulation grows accordingly. Because sRNAs have already been linked many crucial regulatory functions, such as embryonic development, diseases, such as cancer, cardiovascular and neurodegenerative disorders, and response to environmental changes, there is a hope that this better understanding will eventually translate to practical downstream applications in areas such as medical and agricultural research.

Acknowledgments

This work has been funded by the Biotechnology and Biological Sciences Research Council (BBSRC) on supporting Grants: BB/G008078/1 and BB/100016X/1. Also, the authors would like to thank Tamas Dalmay, Simon Moxon, Frank Schwach and Hugh Woolfenden for their valuable guidance and contributions.

References

Addo-Quaye, C., Eshoo, T.W., Bartel, D.P., and Axtell, M.J. (2008). Endogenous siRNA and miRNA targets identified by sequencing of the Arabidopsis degradome. Curr. Biol. *18*, 758–762.

Addo-Quaye, C., Miller, W., and Axtell, M.J. (2009a). CleaveLand: a pipeline for using degradome data to find cleaved small RNA targets. Bioinformatics *25*(1), 130–131.

Addo-Quaye, C., Snyder, J.A., Park, Y.B., Li, Y.F., Sunkar, R., and Axtell, M.J. (2009b). Sliced microRNA targets and precise loop-first processing of MIR319 hairpins revealed by analysis of the Physcomitrella patens degradome. RNA *15*, 2112–2121.

Aho, A.V., and Corasick, M.J. (1975). Efficient String Matching: An Aid to Bibliographic Search. Communications of the ACM *18*, 333–340.

Aird, D., Ross, M.G., Chen, W.-S.S., Danielsson, M., Fennell, T., Russ, C., Jaffe, D.B., Nusbaum, C., and Gnirke, A. (2011). Analyzing and minimizing PCR amplification bias in Illumina sequencing libraries. Genome Biol. *12*, R18+.

Altschul, S.F., Gish, W., Miller, W., Myers, E.W., and Lipman, D.J. (1990). Basic local alignment search tool. J. Mol. Biol. *215*, 403–410.

Andrews, S. (2010). 'FastQC', Babraham Bioinformatics, A quality control tool for high throughput sequence data. Available at: http://www.bioinformatics.bbsrc.ac.uk/projects/fastqc

Axtell, M.J. (2010). A method to discover phased siRNA loci. Methods Mol. Biol. *592*, 59–70.

Axtell, M.J., Jan, C., Rajagopalan, R., and Bartel, D.P. (2006). A two-hit trigger for siRNA biogenesis in plants. Cell *127*, 565–577.

Axtell, M.J., Westholm, J.O., and Lai, E.C. (2011). Vive la différence: biogenesis and evolution of microRNAs in plants and animals. Genome Biol. *12*(4), 221.

Baev, V., Naydenov, M., Apostolova, E., Ivanova, D., Doncheva, S., Minkov, I., and Yahubyan, G. (2010), Identification of RNA-dependent DNA-methylation regulated promoters in Arabidopsis. Plant Physiol. Biochem. *48*, 393–400.

Barrangou, R., Fremaux, C., Deveau, H., Richards, M., Boyaval, P., Moineau, S., Romero, D.A., and Horvath, P. (2007). CRISPR provides acquired resistance against viruses in prokaryotes. Science *315*, 1709–1712.

Benson, G. (1999). Tandem repeats finder: a program to analyze DNA sequences. Nucleic Acids Res. *27*, 573–580.

Bolstad, B.M., Irizarry, R.A., Astrand, M., and Speed, T.P. (2003). A comparison of normalization methods for high density oligonucleotide array data based on variance and bias. Bioinformatics *19*, 185–193.

Bonnet, E., Wuyts, J., Rouze, P., and Van de Peer, Y. (2004). Evidence that microRNA precursors, unlike other non-coding RNAs, have lower folding free energies than random sequences. Bioinformatics *20*, 2911–7.

Bonnet, E., He, Y., Billiau, K., and Van de Peer, Y. (2010). TAPIR, a web server for the prediction of plant microRNA targets, including target mimics. Bioinformatics (Oxford, England) *26*, 1566–1568.

Borsani, O., Zhu, J., Verslues, P.E., Sunkar, R., and Zhu, J.-K. (2005). Endogenous siRNAs derived from a pair of natural cis-antisense transcripts regulate salt tolerance in Arabidopsis. Cell *123*, 1279–1291.

Bullard, J.H., Purdom, E., Hansen, K.D., and Dudoit, S. (2010). Evaluation of statistical methods for normalization and differential expression in mRNA-Seq experiments. BMC Bioinformatics *11*, 94.

Burgyán, J., and Havelda, Z. (2011). Viral suppressors of RNA silencing. Trends Plant Sci. *16*, 265–272.

Burrows, M., and Wheeler, D.J. (1994). A Block-sorting Lossless Data Compression Algorithm', Technical report, Systems Research Center.

Chen, H.-M., Li, Y.-H., and Wu, S.-H. (2007). Bioinformatic prediction and experimental validation of a microRNA-directed tandem trans-acting siRNA cascade in Arabidopsis. Proc. Natl. Acad. Sci. U.S.A. *104*, 3318–3323.

Claverie, J.M., and States, D.J. (1993). Information Enhancement Methods for Large-Scale Sequence-Analysis. Computers and Chemistry *17*(2), 191–201.

Cock, P.J., Fields, C.J., Goto, N., Heuer, M.L., and Rice, P.M. (2010). The Sanger FASTQ file format for sequences with quality scores, and the Solexa/Illumina FASTQ variants. Nucleic Acids Res. *38*, 1767–1771.

Cuperus, J.T., Fahlgren, N., and Carrington, J.C. (2011). Evolution and functional diversification of MIRNA genes. Plant Cell *23*, 431–442.

Dai, X., and Zhao, P.X. (2011). psRNATarget: a plant small RNA target analysis server. Nucleic acids res. *39*(Web Server issue), W155–159.

Darnell, R.B. (2010). HITS-CLIP: panoramic views of protein-RNA regulation in living cells. Wiley Interdiscip. Rev. RNA *1*, 266–286.

Drinnenberg, I.A., Weinberg, D.E., Xie, K.T., Mower, J.P., Wolfe, K.H., Fink, G.R., and Bartel, D.P. (2009). RNAi in budding yeast. Science *326*(5952), 544–550.

Enright, A.J., John, B., Gaul, U., Tuschl, T., Sander, C., and Marks, D.S. (2003). MicroRNA targets in Drosophila. Genome Biol. *5*(1), R1+.

Fahlgren, N., Sullivan, C.M., Kasschau, K.D., Chapman, E.J., Cumbie, J.S., Montgomery, T.A., Gilbert, S.D., Dasenko, M., Backman, T.W.H., Givan, S.A., and Carrington, J.C. (2009). Computational and analytical framework for small RNA profiling by high-throughput sequencing. RNA *15*(5), 992–1002.

Fiume, M., Williams, V., Brook, A., and Brudno, M. (2010). Savant: genome browser for high-throughput sequencing data. Bioinformatics *26*(16), 1938–1944.

Flicek, P., Amode, M.R., Barrell, D., Beal, K., Brent, S., Carvalho-Silva, D., Clapham, P., Coates, G., Fairley, S., Fitzgerald, S., *et al.* (2012). Ensembl 2012. Nucleic Acids Res. *40*(Database issue), D84–90.

Folkes, L., Moxon, S., Woolfenden, H.C., Stocks, M.B., Szittya, G., Dalmay, T., and Moulton, V. (2012). PAREsnip: A Tool for Rapid Genome-Wide Discovery of Small RNA/Target Interactions Evidenced Through Degradome Sequencing. Nucleic Acids Res. *40*(13), e103.

Friedlander, M.R., Chen, W., Adamidi, C., Maaskola, J., Einspanier, R., Knespel, S., and Rajewsky, N. (2008). Discovering microRNAs from deep sequencing data using miRDeep. Nat. Biotech. *26*(4), 407–415.

Gardner, P.P., Daub, J., Tate, J.G., Nawrocki, E.P., Kolbe, D.L., Lindgreen, S., Wilkinson, A.C., Finn, R.D., Griffiths-Jones, S., Eddy, S.R., and Bateman, A. (2009). Rfam: updates to the RNA families database. Nucleic Acids Res. 37(Database issue), D136–140.

Garmire, L.X., and Subramaniam, S. (2012). Evaluation of normalization methods in mammalian microRNA-Seq data. RNA *18*(6), 1279–1288.

Gautier, L., Cope, L., Bolstad, B.M., and Irizarry, R.A. (2004). affy – analysis of Affymetrix GeneChip data at the probe level. Bioinformatics *20*(3), 307–315.

German, M.A., Pillay, M., Jeong, D.H., Hetawal, A., Luo, S., Janardhanan, P., Kannan, V., Rymarquis, L.A., Nobuta, K., German, R., De Paoli, E., Lu, C., Schroth, G., Meyers, B.C., and Green, P.J. (2008). Global identification of microRNA-target RNA pairs by parallel analysis of RNA ends. Nat. Biotechnol. *26*, 941–946.

German, M.A., Luo, S., Schroth, G., Meyers, B.C., and Green, P.J. (2009). Construction of Parallel Analysis of RNA Ends (PARE) libraries for the study of cleaved miRNA targets and the RNA degradome. Nat. Protoc. *4*, 356–362.

Goecks, J., Nekrutenko, A., and Taylor, J. (2010). Galaxy: a comprehensive approach for supporting accessible, reproducible, and transparent computational research in the life sciences. Genome Biol. *11*(8), R86.

Gottwein, E., Mukherjee, N., Sachse, C., Frenzel, C., Majoros, W.H., Chi, J.-T.A., Braich, R., Manoharan, M., Soutschek, J., Ohler, U., and Cullen, B.R. (2007). A viral microRNA functions as an orthologue of cellular miR-155. Nature *450*, 1096–1099.

Greenbaum, D., Colangelo, C., Williams, K., and Gerstein, M. (2003). Comparing protein abundance and mRNA expression levels on a genomic scale. Genome Biol. *4*(9), 117.

Griffiths-Jones, S. (2006). miRBase: the microRNA sequence database. Methods Mol. Biol. *342*, 129–138.

Griffiths-Jones, S. (2010). miRBase: microRNA sequences and annotation. Curr. Protoc. Bioinformatics *12*, 1–10.

Griffiths-Jones, S., Bateman, A., Marshall, M., Khanna, A., and Eddy, S.R. (2003). Rfam: an RNA family database. Nucleic Acids Res. *31*(1), 439–441.

Griffiths-Jones, S., Saini, H.K., van Dongen, S., and Enright, A.J. (2008). miRBase: tools for microRNA genomics. Nucleic Acids Res. *36* (Database issue), D154–158.

Guerra-Assuno, J.A., and Enright, A.J. (2010). MapMi: automated mapping of microRNA loci. BMC Bioinformatics *11*, 133.

Gygi, S.P., Rochon, Y., Franza, B.R., and Aebersold, R. (1999). Correlation between protein and mRNA abundance in yeast. Mol. Cell Biol. *19*, 1720–1730.

Hafner, M., Renwick, N., Brown, M., Mihailović, A., Holoch, D., Lin, C., Pena, J.T.G., Nusbaum, J.D., Morozov, P., Ludwig, J., *et al.* (2011). RNA-ligase-dependent biases in miRNA representation in deep-sequenced small RNA cDNA libraries. RNA *17*, 1697–1712.

Hamilton, A.J., and Baulcombe, D.C. (1999). A species of small antisense RNA in posttranscriptional gene silencing in plants. Science *286*, 950–952.

Hansen, K.D., Brenner, S.E., and Dudoit, S. (2010). Biases in Illumina transcriptome sequencing caused by random hexamer priming. Nucleic Acids Res. *38*, e131.

Hardcastle, T.J. (2012). segmentSeq: methods for identifying small RNA loci from high-throughput sequencing data. Bioconductor version 2.9, R package version 1.6.2.

Hardcastle, T.J., Kelly, K.A., and Baulcombe, D.C. (2012). Identifying small interfering RNA loci from high-throughput sequencing data. Bioinformatics *28*, 457–463.

Heikkinen, L., Kolehmainen, M., and Wong, G. (2011). Prediction of microRNA targets in *Caenorhabditis elegans* using a self-organizing map. Bioinformatics *27*, 1247–1254.

The Heliconius Genome Consortium. (2012). Butterfly genome reveals promiscuous exchange of mimicry adaptations among species. Nature *487*, 94–98.

Herrera, P.R., Ficarra, E., Acquaviva, A., and Macii, E. (2011). miREE: miRNA recognition elements ensemble. BMC Bioinformatics *12*(1), 454+.

Hofacker, I.L. (2003). Vienna RNA secondary structure server. Nucleic Acids Res. *31*, 3429–3431.

Holley, R.W., Apgar, J., Everett, G.A., Madison, J.T., Marquisee, M., Merrill, S.H., Penswick, J.R., and Zamir, A. (1965). Structure of a Ribonucleic Acid. Science *147*, 1462–1465.

Huang, J.C., Babak, T., Corson, T.W., Chua, G., Khan, S., Gallie, B.L., Hughes, T.R., Blencowe, B.J., Frey, B.J., and Morris, Q.D. (2007). Using expression profiling data to identify human microRNA targets. Nat. Methods *4*, 1045–1049.

Huang, P.J., Liu, Y.C., Lee, C.C., Lin, W.C., Gan, R.R., Lyu, P.C., and Tang, P. (2010). DSAP: deep-sequencing small RNA analysis pipeline. Nucleic Acids Res. *38* (Web Server issue), W385–391.

Hubbard, T., Barker, D., Birney, E., Cameron, G., Chen, Y., Clark, L., Cox, T., Cuff, J., Curwen, V., Down, T., *et al.* (2002). The Ensembl genome database project. Nucleic Acids Res. *30*, 38–41.

Huber, W., von Heydebreck, A., Saltmann, H., Poustka, A., and Vingron, M. (2002). Variance stabilization applied to microarray data calibration and to the quantification of differential expression. Bioinformatics *18* (Suppl. 1), S96–104.

Hull, D., Wolstencroft, K., Stevens, R., Goble, C., Pocock, M.R., Li, P., and Oinn, T. (2006). Taverna: a tool for building and running workflows of services. Nucleic Acids Res. *34* (Web Server issue), W729–732.

Humphreys, D.T., Westman, B.J., Martin, D.I.K., and Preiss, T. (2005). MicroRNAs control translation initiation by inhibiting eukaryotic initiation factor 4E/cap and poly(A) tail function. Proc. Natl. Acad. Sci. U.S.A. *102*(47), 16961–16966.

International Human Genome Consortium. (2001). Initial sequencing and analysis of the human genome. Nature *409*(6822), 860–921.

Itaya, A., Bundschuh, R., Archual, A.J., Joung, J.-G., Fei, Z., Dai, X., Zhao, P.X., Tang, Y., Nelson, R.S., and Ding, B. (2008). Small RNAs in tomato fruit and leaf development. Biochim. Biophys. Acta *1779*, 99–107.

Jain, N., Thatte, J., Braciale, T., Ley, K., O'Connell, M., and Lee, J.K. (2003). Local-pooled-error test for identifying differentially expressed genes with a small number of replicated microarrays. Bioinformatics *19*, 1945–1951.

Jansen, R., Embden, J.D.A.V., Gaastra, W., and Schouls, L.M. (2002). Identification of genes that are associated with DNA repeats in prokaryotes. Mol. Microbiol. *43*, 1565–1575.

Jayaprakash, A.D., Jabado, O., Brown, B.D., and Sachidanandam, R. (2011). Identification and remediation of biases in the activity of RNA ligases in small-RNA deep sequencing. Nucleic Acids Res. *39*, e141.

Kawashima, C.G., Yoshimoto, N., Maruyama-Nakashita, A., Tsuchiya, Y.N., Saito, K., Takahashi, H., and Dalmay, T. (2009). Sulphur starvation induces the expression of microRNA-395 and one of its target genes but in different cell types. Plant J. *57*(2), 313–321.

Klus, P., Lam, S., Lyberg, D., Cheung, M.S., Pullan, G., McFarlane, I., Yeo, G.S., and Lam, B.Y. (2012). BarraCUDA – a fast short read sequence aligner using graphics processing units. BMC Res. Notes *5*, 27.

Kolmogorov, A.N. (1965). Three approaches to the quantitative definition of information. Problems of Information Transmission *1*, 1–7.

Kong, Y. (2011). Btrim: a fast, lightweight adapter and quality trimming program for next-generation sequencing technologies. Genomics *98*, 152–153.

Kozomara, A., and Griffiths-Jones, S. (2011). miRBase: integrating microRNA annotation and deep-sequencing data. Nucleic Acids Res. *39* (Database issue), D152–157.

Krek, A., Grün, D., Poy, M.N., Wolf, R., Rosenberg, L., Epstein, E.J., Macmenamin, P., da Piedade, I., Gunsalus, K.C., Stoffel, M., and Rajewsky, N. (2005). Combinatorial microRNA target predictions. Nat. Genet. *37*, 495–500.

Lagos-Quintana, M., Rauhut, R., Lendeckel, W., and Tuschl, T. (2001). Identification of novel genes coding for small expressed RNAs. Science *294*, 853–858.

Langmead, B., Trapnell, C., Pop, M., and Salzberg, S.L. (2009). Ultrafast and memory-efficient alignment of short DNA sequences to the human genome. Genome Biol. *10*, R25.

Lau, N.C., Lim, L.P., Weinstein, E.G., and Bartel, D.P. (2001). An abundant class of tiny RNAs with probable regulatory roles in Caenorhabditis elegans. Science *294*, 858–862.

Le Meur, N., and Gentleman, R. (2012). Analyzing biological data using R: methods for graphs and networks. Methods Mol. Biol. *804*, 343–373.

Lee, R.C., Feinbaum, R.L., and Ambros, V. (1993). The *C. elegans* heterochronic gene lin-4 encodes small RNAs with antisense complementarity to lin-14. Cell *75*, 843–854.

Lewis, B.P., Burge, C.B., and Bartel, D.P. (2005). Conserved seed pairing, often flanked by adenosines, indicates that thousands of human genes are microRNA targets. Cell *120*, 15–20.

Li, F., Orban, R., and Baker, B. (2012). SoMART: a web server for plant miRNA, tasiRNA and target gene analysis. Plant J. *70*, 891–901.

Li, H., and Durbin, R. (2009). Fast and accurate short read alignment with Burrows-Wheeler transform. Bioinformatics *25*, 1754–1760.

Li, H., Ruan, J., and Durbin, R. (2008). Mapping short DNA sequencing reads and calling variants using mapping quality scores. Genome Res. *18*, 1851–1858.

Li, H., Handsaker, B., Wysoker, A., Fennell, T., Ruan, J., Homer, N., Marth, G., Abecasis, G., and Durbin, R. (2009). The Sequence Alignment/Map format and SAMtools. Bioinformatics *25*, 2078–2079.

Li, R., Li, Y., Kristiansen, K., and Wang, J. (2008). SOAP: short oligonucleotide alignment program. Bioinformatics *24*(5), 713–714.

Li, R., Yu, C., Li, Y., Lam, T.W., Yiu, S.M., Kristiansen, K., and Wang, J. (2009). SOAP2: an improved ultrafast tool for short read alignment. Bioinformatics *25*, 1966–1967.

Li, Y.F., Zheng, Y., Addo-Quaye, C., Zhang, L., Saini, A., Jagadeeswaran, G., Axtell, M.J., Zhang, W., and Sunkar, R. (2010). Transcriptome-wide identification of microRNA targets in rice. Plant J. *62*, 742–759.

Linsen, S.E., de Wit, E., Janssens, G., Heater, S., Chapman, L., Parkin, R.K., Fritz, B., Wyman, S.K., de Bruijn, E., Voest, E.E., Kuersten, S., Tewari, M., and Cuppen, E. (2009). Limitations and possibilities of small RNA digital gene expression profiling. Nature Methods *6*, 474–476.

Liu, C.-M., Wong, T., Wu, E., Luo, R., Yiu, S.-M., Li, Y., Wang, B., Yu, C., Chu, X., Zhao, K., Li, R., and Lam, T.-W. (2012). SOAP3: Ultra-fast GPU-based parallel alignment tool for short reads. Bioinformatics *28*, 878–879.

Lopez-Gomollon, S., Mohorianu, I., Szittya, G., Moulton, V., and Dalmay, T. (2012). Diverse correlation patterns between microRNAs and their targets during tomato fruit development indicates different modes of microRNA actions. Planta *236*(6), 1875–1887.

McCormick, K., Willmann, M., and Meyers, B. (2011). Experimental design, preprocessing, normalization and differential expression analysis of small RNA sequencing experiments. Silence 2(1), 2+.

McCue, A.D., Nuthikattu, S., Reeder, S.H., and Slotkin, R.K. (2012). Gene expression and stress response mediated by the epigenetic regulation of a transposable element small RNA. PLoS Genet. *8*(2), e1002474.

MacLean, D., Moulton, V., and Studholme, D.J. (2010). Finding sRNA generative locales from high-throughput sequencing data with NiBLS. BMC Bioinformat. *11*, 93.

Magi, A., Benelli, M., Gozzini, A., Girolami, F., Torricelli, F., and Brandi, M.L. (2010). Bioinformatics for Next-generation Sequencing Data. Genes *1*, 294–307.

Makarova, K.S., Wolf, Y.I., van der Oost, J., and Koonin, E.V. (2009). Prokaryotic homologs of Argonaute proteins are predicted to function as key components of a novel system of defense against mobile genetic elements. Biol. Direct *4*, 29.

Malone, C.D., Brennecke, J., Dus, M., Stark, A., McCombie, W.R., Sachidanandam, R., and Hannon, G.J. (2009). Specialized piRNA pathways act in germline and somatic tissues of the Drosophila ovary. Cell *137*, 522–535.

Mapleson, D., Moxon, S., Dalmay, T., and Moulton, V. (2013). MirPlex: A Tool for Identifying miRNAs in high-throughput sRNA datasets without a genome. J. Exp. Zool. B. Mol. Dev. Evol. *320*(1), 47–56.

Martin, M. (2011). Cutadapt removes adapter sequences from high-throughput sequencing reads. EMBnet. journal, North America.

Millar, A.A., and Waterhouse, P.M. (2005). Plant and animal microRNAs: similarities and differences. Funct. Integr. Genomics 5(3), 129–135.

Mohorianu, I.I. (2012). Deciphering the regulatory mechanisms of sRNAs in plants. PhD thesis, University of East Anglia.

Mohorianu, I., and Moulton, V. (2010). Revealing biological information using data structuring and automated learning. Recent Pat. DNA Gene Seq. *4*, 181–191.

Mohorianu, I., Schwach, F., Jing, R., Lopez-Gomollon, S., Moxon, S., Szittya, G., Sorefan, K., Moulton, V., and Dalmay, T. (2011). Profiling of short RNAs during fleshy fruit development reveals stage-specific sRNAome expression patterns. Plant J. *67*, 232–246.

Mohorianu, I., Lopez-Gomollon, S., Schwach, F., Dalmay, T., and Moulton, V. (2012). FiRePat – Finding Regulatory Patterns between sRNAs and genes. WIRE's Data Mining and Knowledge Discovery 2, 273–284.

Mohorianu, I., Stocks, M.B., Wood, J., Dalmay, T., and Moulton, V. (2013). CoLIde: a bioinformatics tool for CO-expression-based small RNA Loci identification using high-throughput sequencing data. RNA Biology. *10*(7), 1211–1200.

Molnar, A., Schwach, F., Studholme, D.J., Thuenemann, E.C., and Baulcombe, D.C. (2007). miRNAs control gene expression in the single-cell alga Chlamydomonas reinhardtii. Nature *447*(7148), 1126–1129.

Morgulis, A., Gertz, E.M., Schaffer, A.A., and Agarwala, R. (2006). A fast and symmetric DUST implementation to mask low-complexity DNA sequences. J. Comput. Biol. *13*, 1028–1040.

Morin, R., Bainbridge, M., Fejes, A., Hirst, M., Krzywinski, M., Pugh, T., McDonald, H., Varhol, R., Jones, S., and Marra, M. (2008). Profiling the HeLa S3 transcriptome using randomly primed cDNA and massively parallel short-read sequencing. Biotechniques *45*, 81–94.

Mortazavi, A., Williams, B.A., McCue, K., Schaeffer, L., and Wold, B. (2008). Mapping and quantifying mammalian transcriptomes by RNA-Seq. Nat. Methods *5*, 621–628.

Mosher, R.A., Schwach, F., Studholme, D., and Baulcombe, D.C. (2008). PolIVb influences RNA-directed DNA methylation independently of its role in siRNA biogenesis. Proc. Natl. Acad. Sci. U.S.A. *105*, 3145–3150.

Motakis, E.S., Nason, G.P., Fryzlewicz, P., and Rutter, G.A. (2006). Variance stabilization and normalization for one-color microarray data using a data-driven multiscale approach. Bioinformatics *22*, 2547–2553.

Moxon, S., Schwach, F., Dalmay, T., Maclean, D., Studholme, D.J., and Moulton, V. (2008). A toolkit for analysing large-scale plant small RNA datasets. Bioinformatics *24*, 2252–2253.

Murphy, D., Dancis, B., and Brown, J.R. (2008). The evolution of core proteins involved in microRNA biogenesis. BMC Evol. Biol. *8*, 92.

Needleman, S.B., and Wunsch, C.D. (1970). A general method applicable to the search for similarities in the amino acid sequence of two proteins. J. Mol. Biol. *48*, 443–453.

Nicolas, F.E., Moxon, S., de Haro, J.P., Calo, S., Grigoriev, I.V., Torres-Martínez, S., Moulton, V., Ruiz-Vázquez, R.M., and Dalmay, T. (2010). Endogenous short RNAs generated by Dicer 2 and RNA-dependent RNA polymerase 1 regulate mRNAs in the basal fungus Mucor circinelloides. Nucleic Acids Res. *38*, 5535–5541.

Nielsen, J.A., Lau, P., Maric, D., Barker, J.L., and Hudson, L.D. (2009). Integrating microRNA and mRNA expression profiles of neuronal progenitors to identify regulatory networks underlying the onset of cortical neurogenesis. BMC Neurosci. *10*, 98.

Nunez-Iglesias, J., Liu, C.-C., Morgan, T.E., Finch, C.E., and Zhou, X.J. (2010). Joint genome-wide profiling of miRNA and mRNA expression in Alzheimer's disease cortex reveals altered miRNA regulation. PLoS One *5*(2), e8898.

Nygaard, S., Jacobsen, A., Lindow, M., Eriksen, J., Balslev, E., Flyger, H., Tolstrup, N., Moller, S., Krogh, A., and Litman, T. (2009). Identification and analysis of miRNAs in human breast cancer and teratoma samples using deep sequencing. BMC Med. Genomics 2(1), 35+.

Pais, H., Moxon, S., Dalmay, T., and Moulton, V. (2011). Small RNA discovery and characterisation in eukaryotes using high-throughput approaches. Adv. Exp. Med. Biol. 722, 239–254.

Pantaleo, V., Saldarelli, P., Miozzi, L., Giampetruzzi, A., Gisel, A., Moxon, S., Dalmay, T., Bisztray, G., and Burgyan, J. (2010). Deep sequencing analysis of viral short RNAs from an infected Pinot Noir grapevine. Virology *408*(1), 49–56.

Parameswaran, P., Sklan, E., Wilkins, C., Burgon, T., Samuel, M.A., Lu, R., Ansel, K.M., Heissmeyer, V., Einav, S., Jackson, W., *et al.* (2010), Six RNA viruses and forty-one hosts: viral small RNAs and modulation of small RNA repertoires in vertebrate and invertebrate systems. PLoS Pathog. *6*(2), e1000764.

Pasquinelli, A.E., Reinhart, B.J., Slack, F., Martindale, M.Q., Kuroda, M.I., Maller, B., Hayward, D.C., Ball, E.E., Degnan, B., Muller, P., *et al.* (2000). Conservation of the sequence and temporal expression of let-7 heterochronic regulatory RNA. Nature *408*(6808), 86–89.

Pearson, W.R., and Lipman, D.J. (1988). Improved tools for biological sequence comparison. Proc. Natl. Acad. Sci. U.S.A. *85*(8), 2444–2448.

Pelz, C.R., Kulesz-Martin, M., Bagby, G., and Sears, R.C. (2008). Global rank-invariant set normalization (GRSN) to reduce systematic distortions in microarray data. BMC Bioinformatics *9*, 520.

Pepke, S., Wold, B., and Mortazavi, A. (2009). Computation for ChIP-seq and RNA-seq studies. Nat. Methods *6(Suppl 11)*, S22–32.

Peragine, A., Yoshikawa, M., Wu, G., Albrecht, H.L., and Poethig, R.S. (2004). SGS3 and SGS2/SDE1/RDR6 are required for juvenile development and the production of trans-acting siRNAs in *Arabidopsis*. Genes Dev. *18*(19), 2368–2379.

Pfeffer, S., Zavolan, M., Grässer, F.A., Chien, M., Russo, J.J., Ju, J., John, B., Enright, A.J., Marks, D., Sander, C., and Tuschl, T. (2004). 'Identification of virus-encoded microRNAs. Science *304*, 734–736.

Prufer, K., Stenzel, U., Dannemann, M., Green, R.E., Lachmann, M., and Kelso, J. (2008). PatMaN: rapid alignment of short sequences to large databases. Bioinformatics *24*, 1530–1531.

Rajagopalan, R., Vaucheret, H., Trejo, J., and Bartel, D.P. (2006). A diverse and evolutionarily fluid set of microRNAs in Arabidopsis thaliana. Genes Dev. *20*, 3407–3425.

Rajewsky, N. (2006). microRNA target predictions in animals. Nature Genetics *38(Suppl 1(6s))*, S8–13.

Ratcliff, F., Harrison, B.D., and Baulcombe, D.C. (1997). A similarity between viral defense and gene silencing in plants. Science *276*(5318), 1558–1560.

Reczko, M., Maragkakis, M., Alexiou, P., Papadopoulos, G.L., and Hatzigeorgiou, A.G. (2011). Accurate microRNA target prediction using detailed binding site accessibility and machine learning on proteomics data. Frontiers in Genetics 2.

Robinson, M.D., and Oshlack, A. (2010). A scaling normalization method for differential expression analysis of RNA-seq data. Genome Biol. *11*, R25.

Schadt, E.E., Turner, S., and Kasarskis, A. (2010). A window into third-generation sequencing. Hum. Mol. Genet. *19*(R2), R227–240.

Schwab, R., Palatnik, J.F., Riester, M., Schommer, C., Schmid, M., and Weigel, D. (2005). Specific effects of microRNAs on the plant transcriptome. Developmental Cell *8*, 517–527.

Schwach, F., Moxon, S., Moulton, V., and Dalmay, T. (2009). Deciphering the diversity of small RNAs in plants: the long and short of it. Brief Funct. Genomic Proteomic *8*, 472–481.

Sethupathy, P., Megraw, M., and Hatzigeorgiou, A.G. (2006). A guide through present computational approaches for the identification of mammalian microRNA targets. Nature Methods *3*, 881–886.

Seto, A.G., Kingston, R.E., and Lau, N.C. (2007). The coming of age for Piwi proteins. Mol. Cell *26*(5), 603–609.

Shendure, J., and Ji, H. (2008). Next-generation DNA sequencing. Nat. Biotechnol. *26*, 1135–1145.

Siomi, M.C., Sato, K., Pezic, D., and Aravin, A.A. (2011). PIWI-interacting small RNAs: the vanguard of genome defence. Nat. Rev. Mol. Cell Biol. *12*, 246–258.

Smyth, G.K. (2005). Limma: linear models for microarray data. In Bioinformatics and Computational Biology Solutions using R and Bioconductor, Gentleman, R., Carey, V., Dudoit, S., Huber, W., and Irizarry, R. ed (Springer, New York), pp. 397–420.

Smyth, G.K., Yang, Y.H., and Speed, T. (2003). Statistical issues in cDNA microarray data analysis. Methods Mol. Biol. *224*, 111–136.

Sorefan, K., Pais, H., Hall, A., Kozomara, A., Jones, S.G., Moulton, V., and Dalmay, T. (2012). Reducing sequencing bias of small RNAs. Silence 3(1), 4+.

Stern-Ginossar, N., Elefant, N., Zimmermann, A., Wolf, D.G., Saleh, N., Biton, M., Horwitz, E., Prokocimer, Z., Prichard, M., Hahn, G., *et al.* (2007). Host immune system gene targeting by a viral miRNA. Science *317*, 376–381.

Stocks, M.B., Moxon, S., Mapleson, D., Woolfenden, H.C., Mohorianu, I., Folkes, L., Schwach, F., Dalmay, T., and Moulton, V. (2012). The UEA sRNA Workbench: a suite of tools for analysing and visualising next-generation sequencing microRNA and small RNA datasets. Bioinformatics *28*, 2059–2061.

Studholme, D.J. (2012). Deep sequencing of small RNAs in plants: applied bioinformatics. Brief. Funct. Genomics *11*(1), 71–85.

Sturm, M., Hackenberg, M., Langenberger, D., and Frishman, D. (2010). TargetSpy: a supervised machine learning approach for microRNA target prediction. BMC Bioinformatics *11*, 292+.

Sun, G., Wu, X., Wang, J., Li, H., Li, X., Gao, H., Rossi, J., and Yen, Y. (2011). A bias-reducing strategy in profiling small RNAs using Solexa. RNA *17*(12), 2256–2262.

Suo, C., Salim, A., Chia, K.-S., Pawitan, Y., and Calza, S. (2010). Modified least-variant set normalization for miRNA microarray. RNA *16*(, 2293–2303.

Swarts, D.C., Mosterd, C., van Passel, M.W.J., and Brouns, S.J.J. (2012). CRISPR interference directs strand specific spacer acquisition. PLoS One 7(4), e35888.

Takeda, A., Tsukuda, M., Mizumoto, H., Okamoto, K., Kaido, M., Mise, K., and Okuno, T. (2005). A plant

RNA virus suppresses RNA silencing through viral RNA replication. EMBO J. *24*, 3147–3157.

Thomas, M., Lieberman, J., and Lal, A. (2010). Desperately seeking microRNA targets. Nat. Struct. Mol. Biol. *17*, 1169–1174.

Thompson, J.D., Plewniak, F., and Poch, O. (1999). A comprehensive comparison of multiple sequence alignment programs. Nucleic Acids Res. *27*, 2682–2690.

Tian, G., Yin, X., Luo, H., Xu, X., Bolund, L., and Zhang, X. (2010). Sequencing bias: comparison of different protocols of microRNA library construction. BMC Biotechnology *10*, 64+.

The Tomato Genome Consortium. (2012). The tomato genome sequence provides insights into fleshy fruit evolution. Nature *485*, 635–641.

Umbach, J.L., Kramer, M.F., Jurak, I., Karnowski, H.W., Coen, D.M., and Cullen, B.R. (2008). MicroRNAs expressed by herpes simplex virus 1 during latent infection regulate viral mRNAs. Nature *454*, 780–783.

Valencia-Sanchez, M.A., Liu, J., Hannon, G.J., and Parker, R. (2006). Control of translation and mRNA degradation by miRNAs and siRNAs. Genes Dev. *20*, 515–524.

Van der Oost, J., and Brouns, S.J.J. (2009). RNAi: prokaryotes get in on the act. Cell *139*, 863–865.

Vazquez, F., Vaucheret, H., Rajagopalan, R., Lepers, C., Gasciolli, V., Mallory, A.C., Hilbert, J.L., Bartel, D.P., and Crete, P. (2004). Endogenous trans-acting siRNAs regulate the accumulation of *Arabidopsis* mRNAs. Mol. Cell. *16*, 69–79.

Verdel, A., Jia, S., Gerber, S., Sugiyama, T., Gygi, S., Grewal, S.I.S., and Moazed, D. (2004). RNAi-mediated targeting of heterochromatin by the RITS complex. Science *303*, 672–676.

Voinnet, O. (2009). Origin, biogenesis, and activity of plant microRNAs. Cell *136*, 669–687.

Wang, Y.-P., and Li, K.-B. (2009). Correlation of expression profiles between microRNAs and mRNA targets using NCI-60 data. BMC Genomics *10*, 218.

Waterhouse, P.M., Graham, M.W., and Wang, M.B. (1998). Virus resistance and gene silencing in plants can be induced by simultaneous expression of sense and antisense RNA. Proc. Natl. Acad. Sci. U.S.A. *95*, 13959–13964.

Willenbrock, H., Salomon, J., Sokilde, R., Barken, K.B., Hansen, T.N., Nielsen, F.C., Moller, S., and Litman, T. (2009). Quantitative miRNA expression analysis: comparing microarrays with next-generation sequencing. RNA *15*, 2028–2034.

Wu, L., Fan, J., and Belasco, J.G. (2006). MicroRNAs direct rapid deadenylation of mRNA. Proc. Natl. Acad. Sci. U.S.A. *103*, 4034–4039.

Wu, Y., Wei, B., Liu, H., Li, T., and Rayner, S. (2011). MiRPara: a SVM-based software tool for prediction of most probable microRNA coding regions in genome scale sequences. BMC Bioinformatics *12*, 107.

Xie, F., and Zhang, B. (2010). Target-Align: A Tool for Plant microRNA Target Identification. Bioinformatics *26*(23), 3002–3003.

Xie, P., Xiong, Q., Fang, Y., Qing, Q., and Lieber, C.M. (2012). Local electrical potential detection of DNA by nanowire-nanopore sensors. Nat. Nanotechnol. *7*, 119–125.

Xiong, H., Zhang, D., Martyniuk, C.J., Trudeau, V.L., and Xia, X. (2008). Using generalized procrustes analysis (GPA) for normalization of cDNA microarray data. BMC Bioinformatics *9*, 25.

Yang, J.-H., Li, J.-H., Shao, P., Zhou, H., Chen, Y.-Q., and Qu, L.-H. (2010). starBase: a database for exploring microRNA–mRNA interaction maps from Argonaute CLIP-Seq and Degradome-Seq data. Nucleic Acids Res. *39*(Database), D202–209.

Yu, E.Z., Burba, A.E.C., and Gerstein, M. (2007). PARE: a tool for comparing protein abundance and mRNA expression data. BMC Bioinformatics *8*, 309.

Zamore, P.D., Tuschl, T., Sharp, P.A., and Bartel, D.P. (2000). RNAi: double-stranded RNA directs the ATP–dependent cleavage of mRNA at 21 to 23 nucleotide intervals. Cell *101*, 25–33.

Zhang, Y., Wang, X., and Kang, L. (2011). A k-mer scheme to predict piRNAs and characterize locust piRNAs. Bioinformatics *27*(6), 771–776.

Zheng, Y., Li, Y.-F., Sunkar, R., and Zhang, W. (2012). SeqTar: an effective method for identifying microRNA guided cleavage sites from degradome of polyadenylated transcripts in plants. Nucleic Acids Res. *40*(4), e28.

Utility of High-throughput Sequence Data in Rare Variant Detection

5

Viacheslav Y. Fofanov, Tudor Constantin and Heather Koshinsky

Abstract

Rare variant (sequence variants present in < 1% of the sequence data) detection through the analysis of high-throughput sequence (HTS) data is an exciting new field. HTS data enables genotype-based, primer and probe- independent detection of variant sequences that are present in low frequency in the population. This exciting technical advance may find clinical, epidemiological, forensic and quality control applications. However, because the field is new, terminology is not yet consistent, sources of potential error are not characterized, controlled or quantitated, and models to test experimental concepts are not common. Herein, we attempt to (1) provide some terminology framework, (2) review variant detection through traditional approaches and through examination of HTS data, (3) propose a minimum coverage model to test how much HTS data is needed for reliable rare variant detection, (4) examine sources of error that contribute to false positive rare variant detection and (5) discuss potential approaches to minimize these errors.

What is a rare variant?

Variants are individuals that are different than the consensus of the population. In the context of this chapter, the individuals that make up the population (eukaryotes, bacteria, viruses, PCR amplicons, etc.) belong to the same 'species'. In microbial populations, the term 'variant' conveys two distinct concepts: (1) variations within the cells in a population that is sampled at a single time, AND (2) variations that occur amongst distinct strains of the species. This chapter focuses on a sample of a population of one type of organism whose members are homogeneous at the macro level, but where there are DNA sequence differences (variations) present in < 50% of the members in the single sample population. 'Minority', 'rare', and 'ultra-rare' refer to the frequency of the variant. For the purposes of this chapter, the sequence that occurs most frequently in the sample (often occurring in more than 50% of the members of the population) is the dominant variant and is commonly called the reference sequence. Variants that occur in < 50% but > 1% of the population are minority variants. Those that occur in 1.0–0.1% of the population are rare variants and those that occur in < 0.1% of the population are ultra-rare variants. Thus, a rare variant may be a well-described and very common variation that occurs in many of the strains of the organism, BUT in the population of cells in the sample being tested, it occurs in < 1% of the cells.

The possible differences between the dominant (or reference) sequence and the variant (rare, minority or ultra-rare) sequence include single nucleotide polymorphisms (SNPs), multi-nucleotide insertions/deletions, copy number variations, presence/absence of virulence factors, genes, or plasmids, or even large-scale chromosome rearrangements. The most common differences between the reference and variant sequences are SNPs. At each nucleotide position in the reference sequence, there are three possible substitutions, one possible deletion or four possible insertions that need to be considered. Rare SNPs are the SNPs present in the genome of the rare variant. Some tests query a subset of SNPs potentially present in a genomic region while other tests have the potential to query all the SNPs (whether previously known or not) potentially present in a region.

In the clonal propagation of microorganisms, every time a genome replicates there is an opportunity for the daughter genomes to have sequences that are not identical to the parent genome. In virology, the variants are also referred to as quasi-species, and in bacteriology, if the variants were separately isolated, they would be thought of as strains. For simplicity, we will refer to the population as a group of cells from a particular organism. However, the concepts described can equally be applied if the population is virions, amplicons, isolated DNA, enriched DNA, etc.

Why is variant detection needed?

The ability to detect variants (at all frequencies), both known and previously uncharacterized, in bacterial or viral samples has clinical, epidemiological, forensic and quality control applications. In a clinical setting, the detection of rare variants associated with drug resistance that is present in a subpopulation of an individual's clinical sample may inform the treatment decision. This may maximize the chances of a successful outcome and prevent selection of a drug resistant population that was actually present as a rare variant in the population at the time of treatment initiation. In epidemiological studies, the ability to detect rare variants may allow a greater ability to track and to understand the transmission of infectious diseases and the evolution of the viruses/bacteria within the host (Elena and Froissart, 2010). In forensic applications, the pattern of the rare variants could act as a sample's fingerprint and be used in the investigation and prosecution of bioterrorism attacks or attempts. For quality control applications, the comparison of rare variant patterns can be used as a tool for monitoring the genetic consistency of live viral vaccines and other biologicals and for the identification of individual batches (Neverov *et al.*, 2010).

The detection of rare variants is particularly relevant in clinical virology. High mutation rates occur during reverse transcriptase-based replication of RNA viruses and result in a population that contains a wide variety of mutations; the quasi-species. The high genetic diversity allows pre-existing variants in the population to quickly become more abundant under selection pressure. Analysis of deep sequence data of DNA viral genomes shows that, similar to the RNA viruses, DNA viruses consist of complex mixtures (Gorzer *et al.*, 2010; Renzette *et al.*, 2011). The clinical impact of the quasi-species is suspected, but not clear. In HIV (an RNA virus) treatment, an increased risk of not attaining viral suppression has been reported if there are rare variants with resistance mutations present in the patient's viral pool prior to initiation of treatment (Simen, 2009). Some studies support the value of using the knowledge of the rare variants detected to impact treatment protocol (Codoner *et al.*, 2011; Le *et al.*, 2009; Metzner *et al.*, 2009). However, other studies did not find a strong association between the ability to identify rare variants and the clinical response (Jakobsen *et al.*, 2010, Peuchant *et al.*, 2008).

Utility of non-HTS methods for minority and rare variant detection

Variant detection methods can be broadly divided into two groups: phenotype and genotype based. Phenotype-based detection relies on an observable difference in the phenotype and does not require pre-existing knowledge of the genetic sequence. Genotype-based detection relies on differences in the genome that may or may not correspond to a changed phenotype. Genotyping methods not based on sequencing (whole genome or partial genome) only query a subset of SNPs in a region of interest. In contrast, whole-genome or partial genome sequencing can query all the SNPs in a region of interest.

Drug resistance profiles can be determined by cultivating pathogens in the presence and absence of relevant drug concentrations and comparing their growth profiles. When pathogens are grown on solid media containing concentration gradients of drugs, the relative frequency of resistant subpopulations can be detected, sometimes down to 1%. Although these phenotype-based methods provide information that can be directly used in treatment, their disadvantages are that (i) they provide no genetic information and thus cannot be applied to forensic or other situations, (ii) if the pathogen is slow growing (such as *Mycobacterium tuberculosis*), an answer may not be available for weeks, (iii) they produce infectious materials that may be difficult to contain and dispose, and (iv) they can be complicated to perform.

Genotype-based methods are, to date, mainly used to infer drug resistance profiles and require the exact sequences that cause drug resistance to be known. The characteristics of standard genotype-based variant detection methods and their ability to be applied to minority, rare and ultra-rare variation detection are summarized in Table 5.1.

Sanger sequencing can be used to generate the consensus sequence of the region of interest. Manual analysis of the chromatograms can identify minority variants present at >20%. An extension of this Sanger sequence based method of minority variant detection is to PCR amplify the region of interest from hundreds of isolated colonies and generate Sanger sequence data on each amplicon. To be 99% confident of obtaining one or more amplicon of a variant present in 1% of the colonies, Sanger sequence data must be made on >675 amplicons. One approach to overcome this large sample size problem is to use Line probe assays. Line probe assays are based on the hybridization of PCR amplicons to probes immobilized on a surface and are commercially available for the detection of common drug resistance mutations in some pathogens. The hybridization pattern indicates which wild type and mutant sequences are present in the sample. If there is a dedicated probe for a particular variant sequence, then minority variants present at >5–10% can be detected. For mutations not queried by a probe, the reduction in the intensity of the signal of the wild type probe compared to the control probe is usually interpreted as an un-queried mutation present at about 50%. Allele-specific qPCR is the most sensitive method available for the detection of specifically queried ultra-rare variants present at >0.005% (Nasis *et al.*, 2004).

Conceptually the analysis of HTS data for minority, rare and ultra-rare variant detection is similar to the analysis of Sanger sequence data from amplicons for the detection of minority or rare variants. When HTS data is used instead of Sanger sequence data, a lower detection threshold is possible and the cost of generating sequence data is one or two magnitudes lower (Table 5.1). Several studies have shown that the analysis of HTS data allows the detection of low-frequency mutations that are not detectable by analysis of Sanger sequence data (Hoffmann *et al.*, 2007; Simen *et al.*, 2009; Wang *et al.*, 2007).

Status of rare variant detection by analysis of HTS data

Since 2007, sequence data generated on the 454, Illumina and SOLiD platforms have been used in minority, rare and ultra-rare variant detection in applications ranging from identification of HIV quasi-species to cancer cell line phylogeny (Table 5.2). As this is a relatively new field, there is no standard terminology or context for generating and analysing data or discussing results. In some papers, neither the sensitivity nor specificity of the

Table 5.1 Summary of genotype-based variant detection methods

Method	Cost per region of interest[1]	Mutation throughput[2]	All mutations in region of interest	Variant detection threshold	Easy to add new mutations
Sanger	Low	High	Yes	Minority	Yes
Amplicons	High	High	Yes	Rare	Yes
LiPA[3]	Medium	Low–medium	No	Minority	No
qPCR	Low	Low	No	Ultra-rare	No
HTS[4]	Low	Very high	Yes	Ultra-rare	Yes

[1]Cost estimate in mid-2012 for the sequence data generation or query portion of the assay: low, under $20; medium, $21–250; high, >$251. [2]Mutations throughput per experiment: 1–10, low; 11–100, medium; 101– 800, high; >801, very high. [3]Line probe assay. [4]High-throughput sequencing.

Table 5.2 Summary of recent peer reviewed publications describing minority, rare, and ultra-rare variant detection from HTS data

Application	Platform	Year	Level	Sensitivity	Specificity	Error correction	Reference
HIV drug resistance mutations	454	2007	>5% spiked, 0.65%	n.r.	n.r.	Any level above the error rate of the sequencing platform	Hoffmann *et al.* (2007)
Cancer cell lines phylogeny	454	2008	0.02%	n.r.	n.r.	Linked mutations; mutations above error rate of the sequencing platform	Campbell *et al.* (2008)
HBV drug resistance	454	2008	>1%	n.r.	n.r.	1% cut-off was used	Solmone *et al.* (2009)
HIV quasi-species dynamics	454	2010	0.07–0.09%	n.r.	n.r.	Statistically derived cut-off values for detection of all possible mutations at each position based on sequencing of a control plasmid	Hedskog *et al.* (2010)
Fibrous dysplasia diagnosis	454	2011	5%	n.r.	n.r.	None	Liang *et al.* (2011)
HIV-1 *pol* gene variation	454	2011	0.27%	n.r.	n.r.	Bland–Altman analyses and plots	Mild *et al.* (2011)
PCR template molecule counting	454	2011	n.r.	n.r.	n.r.	Template tagging	Casbon *et al.* (2011)
Determining sources of error in sequencing studies	454	2012	1%	n.r.	n.r.	Remove SNPs belonging to homopolymers of length 4 or more	Becker *et al.* (2012)
Sensitive and specific detection of rare variants	454	2012	<1%	>97%	>97%	V-Phaser	Macalalad *et al.* (2012)
Detection of abdominal aortic aneurysm-associated rare variants in human	SOLiD	2011	2.5%	n.r.	n.r.	n.r.	Harakalova *et al.* (2011)
Quantify rare SNPs from pooled genomic DNA (human)	GAI	2009	1.5%	100%	99.8%	SNPSeeker, an in-house developed algorithm based on large deviation theory	Druley *et al.* (2009)
High-throughput discovery of rare variants	GAIIx	2010	0.1%	100%	100%	In-house algorithm	Vallania *et al.* (2010)
Detection and quantification of rare mutations	GAIIx	2011	n.r.	n.r.	n.r.	Individually indexing each DNA template molecule before library prep	Kinde *et al.* (2011)
Rare variant detection in pooled exons samples	GAIIx	2011	5%	n.r.	n.r.	n.r.	Niranjan *et al.* (2011)

n.r., not reported.

rare variant detection is addressed; in others, sensitivity, specificity and positive predictive values are reported, but the definitions do not always match the definitions as used in classical epidemiology. The effect of the variants frequency on the sensitivity and specificity is rarely mentioned. In general one would expect that the lower the frequency, the poorer the sensitivity and specificity would become. While these early papers are largely descriptive, they lay the groundwork for the use of HTS data in rare variant detection.

There are at least three publicly available algorithms for low frequency variant detection. SNPSeeker and SPLINTER are algorithms for SNP detection from Illumina sequence data (Druley *et al.*, 2009; Vallania *et al.*, 2010). V-Phaser is an algorithm for SNP detection in 454 sequence data (Macalalad *et al.*, 2012).

SNPSeeker (Druley *et al.*, 2009) was trained using a control plasmid that was added to 15 PCR amplicons (covering four genes) from 1111 individuals. The sequence data from the first 800 bp of the control plasmid were used to train the algorithm for detecting individual base variations (frequency not indicated). SNPSeeker identified no SNPs in the remaining 467 bp of the control plasmid and in 656 of the 658 bp of one of the amplicons sequenced. Next, SNPSeeker identified all seven SNPs known to be present in a 665 bp amplicon, but also identified two more SNPs that were determined to be false positives as they were not confirmed by qPCR. Fourteen out of 16 SNPs identified from another subset of amplicons were confirmed by qPCR. While 37 of the 55 SNPs that were identified in the remaining three genes were previously described in dbSNP (build 128, http://www.ncbi.nlm.nih.gov/projects/SNP/), the presence of none of the 55 was independently validated.

SPLINTER is an approach that combines specialized library preparation with a computational pipeline for the detection of SNPs and indels from Illumina sequence data at pre-selected loci (Vallania *et al.*, 2010). PCR amplicon pools are mixed with a DNA fragment without variants (negative control), and a synthetic pool with engineered mutations present at the lowest expected variant frequency for the sample (positive control) and sequence data is generated. Analysis of the sequence data of the negative and positive controls, and the actual pooled sample, is used to determine the variants present in the pooled sample. The maximum likelihood approach provides an estimate of the frequency of each variant in the pooled sample. In an initial experiment 15 out of 15 SNPs introduced at frequencies between 0.5% and 0.1% were detected. In a second experiment, 100% sensitivity and between 99.1% and 100% specificity for detecting both SNPs and indels at frequencies from 2% to 0.1%, in a 2260 bp construct was reported. This approach achieves insertion/deletion false positive rates that are comparable with false positive rate for substitutions. SPLINTER has been used to detect rare variants in candidate genes pooled from 439 families at risk for Alzheimer's disease (Cruchaga *et al.*, 2012). While this is a sensitive and specific approach to screen rare variants in selected loci across complex pools of variants present in a wide range of frequencies, it is difficult to extend to multiple loci or whole genome viral, and, more importantly, bacterial and eukaryotic rare variant detection studies, where the more complex genomic features (homopolymers, repeatable regions, orthologous genes, etc.) significantly complicate the rare variant detection process.

V-Phaser is used for SNP detection with 454 sequence data (Macalalad *et al.*, 2012). To illustrate the approach, sequence data were generated from a mixture of eight strains of West Nile virus (10,004 bp). This mixture had 102 variants present at > 2.5%. Analysis of these sequence data for minority variants present at > 2.5% identified 243 variants. Two of the introduced variants were not detected and 143 of the detected variants were false positives. As the frequency of the variant to be detected drops below 2.5%, the detection of false positives, as opposed to actual variants, becomes more pronounced.

Appropriate sample preparation has also been shown to improve rare variant detection. Sequencing errors generated during the amplification step of library preparation can be reduced 70 fold by pre-tagging of each template molecule with a unique barcode. In this way, all library molecules that originated from one initial template can be identified (Kinde *et al.*, 2011; Casbon *et al.*, 2011). As the consensus sequence of each original template molecule can be determined with high confidence, the errors generated during sequence data generation can be detected. This sequencing error control protocol can be applied to any sample type, from chemically synthesized oligos to PCR amplicons to viral and other biological samples.

A recurring theme in minority variant detection by analysis of HTS data is the effort to distinguish true variants from false positives (Zagordi *et al.*, 2010, 2011; Becker *et al.*, 2012; Erlich *et al.*, 2008; Rougemont *et al.*, 2008; Druley *et al.*, 2009; Vallania *et al.*, 2010; Macalalad *et al.*, 2012). It is now generally recognized that the analysis of HTS data for the detection of rare variants requires careful consideration of errors that can be introduced in experimental protocols, from wet lab to bioinformatic analysis (Mild *et al.*, 2011; Vrancken *et al.*, 2010; Wang *et al.*, 2007). The errors must be distinguished from true rare variants. This problem becomes more acute as the variant detection threshold approaches the error rate of the sequencing platform. The sources of errors can be divided into random and systematic. The discussion below focuses on the potential sources of systematic errors and the measurable effects that the correction of these systematic and random errors has on the specificity of variant detection at several thresholds.

How much HTS data is needed to accurately detect rare variants?

A simple statistical model [minimum required coverage (MRC) model] can be used to approximate the average number and length of reads needed to ensure a given coverage of a rare variant (present at a stipulated frequency) with a stipulated degree of certainty. This model can be used to guide the answer to 'How much sequence coverage is required to have at least X coverage of a rare variant present at Y frequency of the sample with Z% certainty?' Assuming that reads are randomly drawn/sequenced from a sample library, the event of obtaining a read from the rare variant can be modelled by independent and identically distributed Bernoulli trials. This means that the coverage of the rare variant (rare SNP used to fingerprint a sample or guide a clinical decision, for example) can be modelled by a binomial (n,p) distribution, where n is the total coverage of the sample and p is the frequency of the rare variant. Thus, the probability of obtaining coverage level, k, for the rare SNP can be described by:

$$P(X=k\,|\,n,p)=\frac{n!}{k!(n-k)!}p^{k}(1-p)^{n-k} \qquad (5.1)$$

From (eqn 5.1), it follows that the probability of obtaining, for example, at least 25× coverage level or higher from the region of interest is given by:

$$1-\sum_{k=0}^{25}P(X=k\,|\,n,p) \qquad (5.2)$$

The model (eqn 5.2) above implies that roughly 19,750× coverage is needed to have 99.0% confidence of obtaining at least 25× coverage for a rare variant present in 0.2% of the library ($P=0.002$). Fig. 5.1 shows the relationship between required total coverage needed, coverage of the rare variant, and frequency of the variant. It should be noted that, in practice, some bias may be observed between the frequency at which a variant is present in the library and the frequency it is present in the biological sample.

The actual number of sequence reads produced from a given location (potential SNP) is not predicted. Rather the model considers the average coverage levels for the entire region of interest (although an average number of reads can be derived from coverage, read length, and length of the region of interest). While other, more complex models (such as those that consider effect of errors within the reads) may produce a better approximation of required minimal coverage, the statistical framework proposed above is a convenient rough approximation that yields conservative estimates in both simulation and real sequencing studies.

This MRC model was tested to determine whether it accurately predicts coverage levels of the rare variant under perfect conditions – *no*

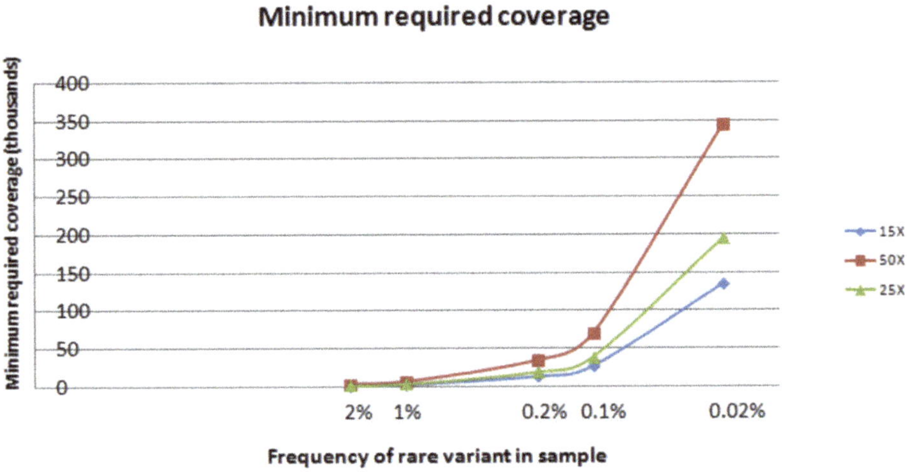

Figure 5.1 Minimum required coverage (in thousands) required in order to be 99.0% confident of obtaining at least 15×, 25×, and 50× coverage of a rare strain present at various frequencies in a sample.

sequencing errors within the reads. A 10 kb region of interest in *Bacillus anthracis* where there are two SNPs that distinguish variant #1 from variant #2 was used. On position 2308 there was a G to A substitution and on position 6622 there was a T to C substitution. Using the reference genome sequence, 50mer reads from variants #1 and #2, without sequencing errors, were computationally generated. Mixtures of reads from variant #2 at a frequency of 2%, 1%, 0.2% and 0.1% in a background of reads from variant #1 were generated. The number of reads generated for each set was determined by the MRC model to attain at least 25× coverage on the rare variant strain with 99.0% confidence. The reads in each mixture were mapped back to the reference sequences of variants #1 and #2. This was repeated 100 times for each of the four mixtures (Table 5.3).

Two conclusions can be drawn from this set of simulations. First, the median coverage of the rare variant is above the predicted 25× threshold. This is not surprising since in the model 25× coverage is exceeded 99% of the time. Second, while there were no positions within the 10 kb reference of the rare variant strain with zero coverage, there were positions with less than 25× coverage. This is also not surprising; 1% of positions are expected to fall below the 25× coverage. It appears that these regions of 'low' coverage were relatively small, as there were 0 experiments out of 100 (for all four mixtures) where 5% of the 10 kb reference sequence had less than 25× coverage.

Table 5.3 Summary of simulations of minority and rare variant coverage using simulated reads without errors

			Variant #1 (statistics averaged over 100 simulations)		Variant #2 (rare variant) (statistics averaged over 100 simulations)			
Expected rare variant coverage	Frequency	Reads generated	Median coverage	Minimum coverage	Median coverage	Minimum Coverage	Experiments with first percentile <25×	Experiments with fifth percentile <25×
25×	2%	380,000	1872	1728	37.37	18.45	65	0
25×	1%	780,000	3880	3673	38.73	18.89	23	0
25×	0.2%	3,900,000	19,560	19,088	38.99	20.09	15	0
25×	0.1%	7,800,000	38,483	39,161	39.12	20.04	13	0

The MRC model does not take into consideration the error rate of the sequence data generation. These errors occur at all steps in the process – from the generation of the primary amplicons to the base calls of the instrument. Errors within the sequence read may cause a reduction in reads available for mapping (when reads cannot be mapped to the reference genome within a stipulated mismatch tolerance level), or may result in false positive rare SNP calls. Another set of simulations was performed using simulated reads with errors added. In Illumina GAIIx sequence data, the location of errors within the read is non-random; errors accumulate towards the ends of the read (Dohm *et al.*, 2008). The spatial distribution of errors identified in sequence data that we generated from 24 libraries of *Enterobacteria phage phiX174* (NC_001422) is shown in Fig. 5.2.

Using the same 10 kb region of *B. anthracis*, 50-mer reads from variants #1 and #2 with sequencing errors (at 0.5% with insertions, deletions and all substitutions being equally likely) in the spatial pattern outlined in Fig. 5.2 were generated *in silico*. Mixtures of these reads from variant #2 at 2%, 1% and 0.2% in a background of variant #1 were generated. The number of reads generated for each set was determined by the model of coverage prediction. The reads in each mixture were mapped back to the reference sequences of variants #1 and #2 and the number of reads that mapped at the two positions of interest perfectly and with mismatches was recorded. This was repeated 100 times for each of the three mixtures (Table 5.4).

Three conclusions can be made from this set of experiments. First, the median coverage levels for variant #1 and variant #2 are consistent with those reported in Table 5.3 – there are no significant changes in the behaviour of the MRC model when errors are added. The error rates modelled in this simulation do not significantly impact the coverage of either the dominant or rare variants. Second, the sequencing errors in reads from the rare variant do not significantly impact our rare SNP detection. The number of reads from variant #2 that map perfectly to the two locations of interest are between 35 and 38, with, on average, less than 0.46 of these reads exhibiting errors. This implies that the errors within variant #2 reads will not be a major source of potentially false positive SNPs calls. Third, past a certain frequency, the sequencing errors from variant #1 will result in false positive variant calls of rare SNP detection. According to the simulation, at frequencies below 1%, the number of reads with errors from the dominant strain (variant #1) exceed the number

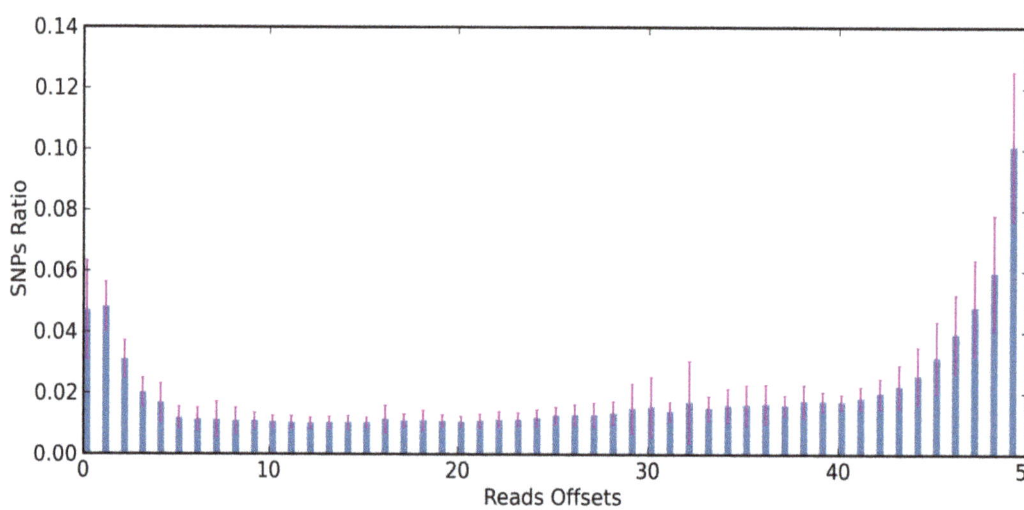

Figure 5.2 Spatial probability mass distribution of errors observed in Illumina GAIIx sequence data we generated from 24 libraries of *Enterobacteria phage phiX174* (NC_001422). Positions of the errors within the sequence read are shown on the x-axis, while the average error rate (across all 24 libraries) and the estimates of standard deviations of errors are shown on the y-axis.

Table 5.4 Summary of simulations of rare variant coverage (calibrated to attain at least 25× coverage of the minority and rare strain) using simulated reads with errors

Frequency	Median coverage	Minimum coverage	Reads mapped perfectly	Reads mapped with errors	Average no. of errors/perfect matches at position 2308	Average no. of errors/perfect matches at position 6622
Variant #1						
2%	1859	1712	369,853	10,322	13.00/1,858	9.00/1,905
1%	3834	3618	766,716	21,335	37.40/3,834	27.20/3,870
0.2%	19,431	18,254	3,864,284	107,576	187.50/18,578	133.77/18,782
Variant #2						
2%	36.93	18.09	7396	204	0.46/38	0.33/35
1%	38.5	19.8	7689	215	0.40/37	0.24/38
0.2%	38.8	19.72	7753	210	0.41/38	0.32/36

of reads confirming the rare SNPs from variant #2, resulting in a false positive call. For example, at position 2308 an average of 37.4 reads from variant #1 mapped (false positives) and 37 reads from variant #2 mapped (confirming the rare SNP). This is not surprising, given the error rate in the simulation. This does not mean that rare variant detection at lower frequencies is impossible, but rather that beyond this point additional measures need to be taken to distinguish sequencing errors from rare variant reads.

The above *in silico* simulations suggest that the MRC model provides an acceptable rough estimate of the total amount of sequence data that must be generated to obtain a stipulated coverage level of the rare variant strain present at a given frequency, with a stipulated confidence. An important practical consideration is the cost to obtain the desired coverage. Specifically, 'How much sequence data would be needed to perform whole-genome rare variant detection in a population of bacterial cells at a given frequency?'

As of May 2012, the specification for an Illumina Genome Analyser II was the production of up to 40 million reads per lane (http://www.illumina.com/systems/genome_analyzer_iix/performance_specifications.ilmn, 15 May 2012). With paired-end sequence reads of 50 b in length, this constitutes 800× coverage of a 5 Mb bacterial genome. Thus, sequence data from more than 24 lanes of paired-end 50 cycle sequencing would be required to have 99.0% confidence to have at least 25× coverage of the rare SNP that was present in 0.2% of the microorganisms in the sample. The situations where such an expenditure of time and money is appropriate are likely limited. In most cases, variants in only a portion of the genome (regions of interest) need to be determined. The regions of interest differ with the application. For rare variant drug resistance detection, the regions of interest would be those involved in drug resistance. For epidemiological, forensic or QC applications, the regions of interest would be known hypervariable regions.

Testing the feasibility of analysing HTS for rare SNP detection

The MRC model outlined above is naive to actual samples: the model assumes that *all* the nucleic acid is high quality and comes from the microorganism of interest. This would occur with cultured microorganisms, but is unlikely to occur with primary forensic (field-acquired) or clinical samples.

The work of others (Table 5.2) has shown that HTS data generation on enriched regions of interest for variant detection has the potential to resolve the issues of (i) DNA from organisms not of interest, and (ii) low quantity DNA. As the specificity of amplification reactions increases, the amount of DNA for sequencing from the region of interest, the expense of generating sequence data from non-informative regions or from other organisms present in the sample is reduced. As the amplification reactions will proceed with limited amounts of starting DNA, the issue of low quantity of primary material is also minimized. Thus, for most applications the addition of an amplification enrichment step reduces the required amount of sequence data to a number that can be accessibly generated with platforms that are currently available.

According to the MRC model, substantial coverage is required to guarantee that a sufficient amount of sequence data is generated from the rare variant genome or region of interest and that distinguishing true rare variant SNPs from noise (false positives) is likely to be a major limiting factor of the use of HTS data for the detection of rare variants. The feasibility of rare variant detection in a pool of PCR amplified sequences was experimentally tested. PCR-amplified segments of a microbial genome were chosen as these represent the likely source of materials in most rare variant detection applications. Additionally, pools with specific base changes can be generated and mixed at required frequencies.

Six short (3.5 kbp total), PCR amplified segment of DNA from *Mycobacterium tuberculosis* were produced as stocks and used to make mixtures. One amplicon was generated in two forms: (i) wild type and (ii) with a mutation (substitution of A to G on position 150). In addition to the pools of 100% wild-type amplicons and of 100% mutant amplicons, two mixtures of amplicons at 0.1% mutant in a background of wild type were prepared (the actual observed frequency of the mutant was 0.2%). Standard Illumina and Nextera based libraries were created and single-end 36 cycle sequence data was generated on an Illumina GAIIx. The sequence data were mapped to the PCR amplicon 'reference genome' (Table 5.5) and the SNPs at various detection thresholds were identified (Table 5.6).

At 0.1% detection threshold (all variants present at frequency of 0.1% or greater), the true positive SNP was identified in each sample mixture. However, 459 additional and likely false positive SNPs were also detected in mixture #1 (Nextera library prep) and 242 additional and likely false positive SNPs were detected in mixture #2 (standard Illumina library prep). When no correction methods are used, the rate of false positive SNP detection ranges from an average of 1 per kb of amplicon at the 1% detection threshold to an average of 100 per kb of amplicon at the 0.1% detection threshold (Table 5.7). These false-positive rare SNPs would not be a problem in a typical dominant SNP identification setting – SNPs observed at < 1% are typically below detection thresholds for routine SNP detection studies (usually > 50%), and as such do not contribute to the false positive rate. However, in rare

Table 5.5 Excerpt from mapping summary for standard Illumina 0.1% mixture experiment (mixture #2). The number of reads supporting a perfect match (PM) or each type of SNP are indicated. The true positive SNP is on position 150 (Sub G)

Position	Base	PM	Sub-A	Sub-T	Sub-G	Sub-C	Del	Ins-A	Ins-T	Ins-G	Ins-C
149	C	176,659	115	60	23	0	16	9	6	6	5
150	A	176,893	0	27	356	53	7	9	6	11	3
151	C	177,704	35	57	37	0	5	6	2	13	3

Table 5.6 Summary of rare variant SNPs detected at four detection thresholds, before and after the application of correction methods to sequence data obtained from amplicons (3.5kb total length) of *M. tuberculosis*

	0.1%	0.2%	0.5%	1%
Mixture #1 (Nextera library)	460	119	11	3
Mixture #2 (standard Illumina library)	243	36	10	5
Mixture #1 after spatial correction	22	4	2	2
Mixture #2 after spatial correction	98	16	5	3
Common after spatial correction in both mixtures	13	3	2	2

Table 5.7 Average (mixture #1 and mixture #2) number of false positive rare variant SNPs detected per kb of amplicon at four detection thresholds

	0.1%	0.2%	0.5%	1%
No correction	100.1	21.9	2.7	0.9
Spatial correction	16.9	2.6	0.7	0.4
Common after spatial correction	3.4	0.6	0.3	0.3

variant detection, when the detection threshold is necessarily low and the signal-to-noise ratio is low, the elimination of false positives is critical.

The lower detection threshold, and by extension, the expected sensitivity is anticipated to not be a limitation of the system. When sequence data generated from an amplicon mixture with a SNP added in at 0.01% are analysed, the true positive SNP is routinely detected (data not shown). Numerous additional, and likely false positive, SNPs are also detected.

Sources of errors

Detailed examination of identified false-positive SNPs revealed three sources of systematic errors: (1) sequencing platform errors, (2) errors associated with reference sequence features and (3) alignment algorithm associated errors. Quantifying and minimizing these systematic sources of error is therefore critical before HTS data can routinely and robustly be used in rare variant detection.

The sequencing platform related sequencing errors in Illumina GAIIx reads exhibit a well documented spatial distribution (Dohm *et al.*, 2008) along the length of the read (Fig. 5.2). The quality of the base-calls produced by the sequencer (in the form of Phred quality scores, which are logarithmically related to the error probability of each base call) can provide useful insights into the expected error rates for each base in the sequencing read. While the correlation between expected error rates (as per Phred score) and actual error rates (when mapped to known reference genome) is very good (Fig. 5.3), application of the quality scores to rare variant prediction (quality scores corresponding to error rates of < 0.1%) is difficult. In practice, removal of reads with one or more low quality base scores constitutes a significant drop (40–80% in some extreme cases) in available reads (data not shown).

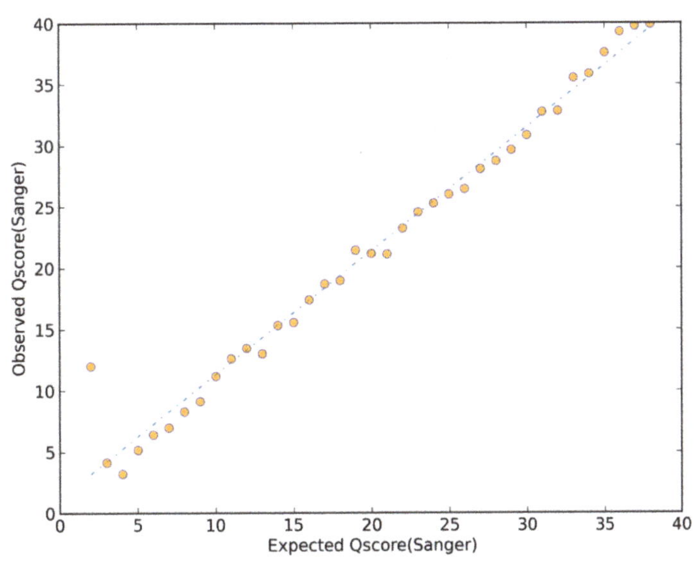

Figure 5.3 Expected quality scores (in Sanger Phred format) produced by Illumina GAIIx sequencing platform (*Pseudomonas aeruginosa* (NC_008463) sample) versus observed quality scores that we obtained by mapping the sequence reads (with up to 3 mismatches) to the reference genome using BWA mapping algorithm.

Features of the reference sequence also can contribute to false positive rare SNP detection. Repeatable regions, low complexity regions, and homopolymers, all have higher rates of false positive SNP detection. Repeatable subsequences or nearly repeatable subsequences (those repeating with one mismatch elsewhere) within the reference genome will result in ambiguity when attributing a rare SNP. Homopolymer subsequences (e.g. AAAA) and nearly homopolymer (e.g. CAC, ACC or CAA) subsequences are more prone to sequencing errors than other subsequences. We have observed in phiX samples that sequencing errors which result in a homopolymer gain (e.g. CAC to CCC) are approximately two times more likely than those that result in homopolymer loss (e.g. CCC to CAC) or are homopolymer neutral (data not shown). Mapping algorithms themselves can introduce errors due to ambiguous or incorrect alignment of short sequence reads onto the reference sequence of the dominant strain in the sample. Fortunately, these systematic sources of error are likely correctable.

There is also a random component to false positive SNP identification. This type of error is typical of multiple hypothesis-testing problems that arise when each of the bases in the reference genome is queried to determine whether it contains a SNP. In all, eight independent hypothesis tests are done for each position (three substitutions, one deletion and four insertions). The large number of tests combined with the requirement for low SNP detection threshold necessary for rare variant detection, leads to some positions being labelled as SNPs due to random chance. In addition, errors produced by the polymerase in the library preparation steps will be largely random in nature.

Experimental validation of correction approaches

Though other methods may be possible, we have approached the problem of improving the specificity of rare variant detection from the point of view of controlling two of the main sources of errors – systematic sequencing platform errors and random sequencing error noise.

The systematic sequencing platform errors were addressed by leveraging the spatial distribution of errors along the read to help in SNP identification process; the mismatches within X bases (in our case ranging from 1 to 5) from the edges of the reads, were not allowed to contribute to the SNP calling process. Since the edges of the sequence reads are the most prone-to-error portion of the read (Fig. 5.2) and are also more difficult to properly align, this should have a significant false-positive reduction effect. Another option to minimize the random errors is to generate multiple replicates of the sequence data and only consider SNPs that are detected in all replicates as true SNPs. The random SNPs are unlikely to appear in the same position twice.

These two correction approaches were applied to the sequence data generated from the mixtures of the six amplicons from *M. tuberculosis* described in section 'Testing the feasibility of analysing HTS for rare SNP detection' (Table 5.6). At the 0.1% detection threshold, the spatial correction alone ($X=5$, freezing 5 bases from each edge of the 36-nt-long read) reduced the number of SNPs detected from 460 to 22 (mixture 1) and from 243 to 98 (mixture 2). If only the SNPs that are present in both of the mixtures (after spatial correction) are considered, the number of SNPs is further reduced to 13. The true positive SNP (A>G at position 150 in one amplicon) was detected in both mixtures, before and after the application of correction procedures.

At the 0.2% detection threshold the spatial correction reduced the number of SNPs detected from 119 to 4 (mixture 1) and from 36 to 16 (mixture 2). Three SNPs were common to both sets of sequence data, one of which is the true positive SNP. At the 0.5% detection threshold the spatial correction alone reduced the number of SNPs detected from 11 to 2 (mixture 1) and from 10 to 5 (mixture 2) and two SNPs were common to both sets of sequence data.

In all cases, the true positive SNP (A>G at position 150 in one amplicon) was detected in both mixtures (at 0.1% and 0.2% detection thresholds). However, an additional SNP was consistently detected as being present in 1.18% of the sequence data. As this minority SNP was not intentionally added to the amplicon pool it is considered a false positive. It is either present in the biological sample or an error produced during amplicon generation and underscores the need to have the replicates include the amplicon (or other enrichment) generation step.

On this data at the 1% detection threshold, spatial correction alone is sufficient to reduce the false positive SNPs to one or two over the 3.5 kb region. Furthermore, spatial correction combined with technical replicates is sufficient to significantly remove false positive SNPs on this dataset at the 0.2% detection threshold.

Conclusions

Simulation and experimental data suggest that HTS approaches can affordably achieve the depth of coverage necessary for sensitive and specific detection of rare variant SNPs even at the 0.1% frequency (Table 5.6). The literature on computational simulations and experimental validations shows that the detection of false positive SNPs present the largest challenge to sensitive and specific rare variant identification. At least three systematic sources of error are present: (1) sequencing platform-related sequencing error, including spatial distribution of errors and homopolymer biases; (2) reference sequence-induced bias; (3) alignment algorithm bias. There is also a random component of errors (related to the multiple hypothesis testing problem and polymerase errors) that contribute to false positive rare SNP detection. There are steps to mitigate the impact of each of these errors (Table 5.8). Addressing systematic and random sources of errors requires development of specialized mapping and SNP detection approaches. We anticipate that these errors may be removed by a combination of improving mapping and SNP detection algorithms, annotating the reference genome for areas prone to false positive SNP detection and increasing the accuracy of the data by replicating the sequence data generated.

The experimental results indicate that the application of spatial correction and technical replicates (as appropriate to the detection threshold needed for the application) increased the specificity of minority and rare SNP detection. As ultra-rare SNPs can be detected, the sensitivity of rare SNP detection is unlikely to be impacted. False positive SNP detection was reduced and no new false positive SNPs were detected. If detection of a rare variant present at a frequency of <0.1% is required, further library preparation and computational tools are needed. These tools are in development and include specialized mapping and reference genome annotation (Table 5.8).

Table 5.8 Summary of the sources of error in rare variant detection and potential fixes

Source of Error	Fix
Systemic	
Sequencing platform	Specialized library preparation protocols
Reference sequence	Annotating reference genome
Alignment method	Specialized mapping
Random	Replicate sequence data generation and analysis

Acknowledgements

We would like to acknowledge M. Shin for library preparation and sequence data generation. This work was supported in part through DHS Contract N10PC20199.

References

Becker, E.A., Burns, C.M., León, E.J., Rajabojan, S., Friedman, R., Friedrich, T.C., O'Connor, S.L., and Hughes, A.L. (2012). Experimental analysis of sources of error in evolutionary studies based on roche/454 pyrosequencing of viral genomes. Genome Biol. Evol. 4,457–465.

Campbell, P.J., Pleasance, E.D., Stephens, P.J., Dicks, E., Rance, R., Goodhead, I., Follows, G.A., Green, A.R., Futreal, P.A., Stratton, and M.R. (2008). Subclonal phylogenetic structures in cancer revealed by ultra-deep sequencing. PNAS. *105,* 13081–13086.

Casbon, J.A., Osborne, R.J., Brenner, S., and Lichtenstein, C.P. A method for counting PCR template molecules with application to next-generation sequencing. (2011). Nucleic Acids Res. *39,* e81.

Codoñer, F.M., Pou, C., Thielen, A., García, F., Delgado, R., Dalmau, D., Álvarez-Tejado, M., Ruiz, L., Clotet, B., and Paredes, R. (2011). Added value of deep sequencing relative to population sequencing in heavily pre-treated HIV-1-infected subjects. PLoS One *6,* e19461.

Cruchaga, C., Chakraverty, S., Mayo, K., Vallania, F.L., Mitra, R.D., Faber, K., Williamson, J., Bird, T., Diaz-Arrastia, R., Foroud, *et al.* (2012). Rare variants in APP, PSEN1 and PSEN2 increase risk for AD in late-onset Alzheimer's disease families. PLoS One 7, e31039.

Dohm, J.C., Lottaz, C., Borodina, T., and Himmelbauer, H. (2008). Substantial biases in ultra-short read datasets from high-throughput DNA sequencing. Nucleic Acids Res. *36,* e105.

Druley, T.E., Vallania, F.L., Wegner, D.J., Varley, K.E., Knowles, O.L., Bonds, J.A., Robison, S.W., Doniger, S.W., Hamvas, A., Cole, F.S., Fay, J.C., and Mitra, R.D. (2009). Quantification of rare allelic variants from pooled genomic DNA. Nat. Methods *6,* 263–265.

Elena, S.F., and Froissart, R. (2010). New experimental and theoretical approaches towards the understanding of the emergence of viral infections. Introduction. Philos. Trans. R. Soc. Lond. B. Biol. Sci. *27,* 1867–1869.

Erlich, Y., Mitra, P.P., delaBastide, M., McCombie, W.R., and Hannon, G.J. (2008). Alta-Cyclic: a self-optimizing base caller for next-generation sequencing. Nat. Methods *5,* 679–682.

Görzer, I., Guelly, C., Trajanoski, S., and Puchhammer-Stöckl, E. (2010). Deep sequencing reveals highly complex dynamics of human cytomegalovirus genotypes in transplant patients over time. J. Virol. *84,* 7195–7203.

Harakalova, M., Nijman, I.J., Medic, J., Mokry, M., Renkens, I., Blankensteijn, J.D., Kloosterman, W., Baas, A.F., and Cuppen, E. (2011). Genomic DNA pooling strategy for next-generation sequencing-based rare variant discovery in abdominal aortic aneurysm regions of interest-challenges and limitations. J. Cardiovasc. Transl. Res. *4,* 271–280.

Hedskog, C., Mild, M., Jernberg, J., Sherwood, E., Bratt, G., Leitner, T., Lundeberg, J., and Andersson, B. (2010). Albert J. Dynamics of HIV-1 quasispecies during antiviral treatment dissected using ultra-deep pyrosequencing. PLoS One *5,* e11345.

Hoffmann, C., Minkah, N., Leipzig, J., Wang, G., Arens, M.Q., Tebas, P., and Bushman, F.D. (2007). DNA bar coding and pyrosequencing to identify rare HIV drug resistance mutations. Nucleic Acids Res. *35,* e91.

Jakobsen, M.R., Tolstrup, M., Søgaard, O.S., Jørgensen, L.B., Gorry, P.R., Laursen, A., and Ostergaard, L. (2010). Transmission of HIV-1 drug-resistant variants: prevalence and effect on treatment outcome. Clin. Infect. Dis. *50,* 566–573.

Kinde, I., Wu, J., Papadopoulos, N., Kinzler, K.W., and Vogelstein, B. (2011). Detection and quantification of rare mutations with massively parallel sequencing. PNAS *108,* 9530–9535.

Le, T., Chiarella, J., Simen, B.B., Hanczaruk, B., Egholm, M., Landry, M.L., Dieckhaus, K., Rosen, M.I., and Kozal, M.J. (2009). Low-abundance HIV drug-resistant viral variants in treatment-experienced persons correlate with historical antiretroviral use. PLoS One *4,* e6079.

Liang, Q., Wei, M., Hodge, L., Fanburg-Smith, J.C., Nelson, A., Miettinen, M., Foss, R.D., and Wang, G. (2011). Quantitative analysis of activating alpha subunit of the G protein (Gsα) mutation by pyrosequencing in fibrous dysplasia and other bone lesions. J. Mol. Diagn. *13,* 137–42.

Macalalad, A.R., Zody, M.C., Charlebois, P., Lennon, N.J., Newman, R.M., Malboeuf, C.M., Ryan, E.M., Boutwell, C.L., Power, K.A., Brackney, D.E., *et al.* (2012). Highly sensitive and specific detection of rare variants in mixed viral populations from massively parallel sequence data. PLoS Comput. Biol. *8,* e1002417.

Metzner, K.J., Giulieri, S.G., Knoepfel, S.A., Rauch, P., Burgisser, P., Yerly, S., Günthard, H.F., and Cavassini, M. (2009). Minority quasispecies of drug-resistant HIV-1 that lead to early therapy failure in treatment-naive and -adherent patients. Clin. Infect. Dis. *48,* 239–247.

Mild, M., Hedskog, C., Jernberg, J., and Albert, J. (2011). Performance of ultra-deep pyrosequencing in analysis of HIV-1 pol gene variation. PLoS One *6,* e22741.

Nasis, O., Thompson, S., Hong, T., Sherwood, M., Radcliffe, S., Jackson, L., and Otevrel, T. (2004). Improvement in sensitivity of allele-specific PCR facilitates reliable noninvasive prenatal detection of cystic fibrosis. Clin. Chem. *50,* 694–701.

Neverov, A., and Chumakov, K. (2010). Massively parallel sequencing for monitoring genetic consistency and quality control of live viral vaccines. Proc. Natl. Acad. Sci. U.S.A. *107,* 20063–20068.

Niranjan, T.S., Adamczyk, A., Bravo, H.C., Taub, M.A., Wheelan, S.J., Irizarry, R., and Wang, T. (2011). Effective detection of rare variants in pooled DNA samples using Cross-pool tailcurve analysis. Genome Biol. *12,* R93.

Peuchant, O., Thiébaut, R., Capdepont, S., Lavignolle-Aurillac, V., Neau, D., Morlat, P., Dabis, F., Fleury, H., Masquelier, B., and ANRS CO3 Aquitaine Cohort. (2008). Transmission of HIV-1 minority-resistant variants and response to first-line antiretroviral therapy. AIDS *22,* 1417–14123.

Renzette, N., Bhattacharjee, B., Jensen, J.D., Gibson, L., and Kowalik, T.F. (2011). Extensive genome-wide variability of human cytomegalovirus in congenitally infected infants. PLoS Pathog. *7,* e1001344.

Rougemont, J., Amzallag, A., Iseli, C., Farinelli, L., Xenarios, I., and Naef, F. (2008). Probabilistic base calling of Solexa sequencing data. BMC Bioinformatics *9,* 431.

Simen, B.B., Simons, J.F., Hullsiek, K.H., Novak, R.M., Macarthur, R.D., Baxter, J.D., Huang, C., Lubeski, C., Turenchalk, G.S., Braverman, M.S., *et al.* (2009). Terry Beirn Community Programs for Clinical Research on AIDS, Low-abundance drug-resistant viral variants in chronically HIV-infected, antiretroviral treatment-naive patients significantly impact treatment outcomes. J. Infect. Dis. *199,* 693–701.

Solmone, M., Vincenti, D., Prosperi, M.C., Bruselles, A., Ippolito, G., and Capobianchi, M.R. (2009). Use of massively parallel ultradeep pyrosequencing to characterize the genetic diversity of hepatitis B virus in drug-resistant and drug-naive patients and to detect minor variants in reverse transcriptase and hepatitis B S antigen. J. Virol. *83,* 1718–1726.

Vallania, F.L., Druley, T.E., Ramos, E., Wang, J., Borecki, I., Province, M., and Mitra, R.D. (2010). High-throughput discovery of rare insertions and deletions in large cohorts. Genome Res. *20,*1711–1718.

Vrancken, B., Lequime, S., Theys, K., and Lemey, P. (2010). Covering all bases in HIV research: unveiling a hidden world of viral evolution. AIDS Rev. *12*, 89–102.

Wang, C., Mitsuya, Y., Gharizadeh, B., Ronaghi, M., and Shafer, R.W. (2007). Characterization of mutation spectra with ultra-deep pyrosequencing: application to HIV-1 drug resistance. Genome Res. *17*, 1195–1201.

Zagordi, O., Klein, R., Däumer, M., and Beerenwinkel, N. (2010). Error correction of next-generation sequencing data and reliable estimation of HIV quasispecies. Nucleic Acids Res. *38*, 7400–7409.

Zagordi, O., Bhattacharya, A., Eriksson, N., and Beerenwinkel, N. (2011). Estimating the genetic diversity of a mixed sample from next-generation sequencing data. BMC Bioinformatics *12*, 119.

Detecting Breakpoints of Insertions and Deletions from Paired-end Short Reads

6

Kai Ye and Zemin Ning

Abstract

There is a strong demand in the genomics community to develop effective algorithms to reliably identify all types of genomic variants. Indel and structure variant detection using next-generation sequencing (NGS) data is difficult, and identification of large and complex structural variations is extremely challenging. When applied to NGS data, split-read methods have recently demonstrated their power both in pinpointing the precise breakpoints and in efficient use of computer memory and time, as compared with the read-depth, read-pair and assembly approaches. Pindel and its recent improvements as well as other split-read approaches are reviewed in this article. As each current method can only capture a subset of variant types with a high degree of confidence, a complete software package is needed in the field in order to integrate all types of signals for identifying all genetic variants of different types and size ranges.

Introduction

Each individual has a unique genomic makeup. The complete catalogue of genetic variants contains single base substitutions or single nucleotide polymorphisms (SNP), small insertions and deletions, and large structural variants. With recent progresses on sequencing technologies and computational methodologies, it has become apparent that non-SNP variants, especially large chromosomal rearrangement events such as structural variants, define the majority of inter-individual differences. In previous studies, hundreds of large-scale structural variants have been identified using arrayCGH (Iafrate *et al.*, 2004). Polymorphic transposon insertions (Bennett, *et al.*, 2004) and other short indel polymorphisms (Mills *et al.*, 2006) have been discovered using capillary sequencing reads. Clone-end capillary sequencing provided a means to interrogate the intermediate size structural variations and, in particular, the method is able to find relatively large insertion events (Kidd *et al.*, 2008). The complete sequencing of whole genomes can create detailed catalogues of structural variations within and between human individuals (Levy *et al.*, 2007). For example, the de novo assembly of the Venter genome played an essential role in detecting long insertions, deletions and structural rearrangements.

Recent efforts on variant detection have relied on NGS high-throughput data. Using the Illumina platform for a complete sequencing of a human individual, Bentley *et al.* (2008) reported ~4-million SNPs and ~0.4-million short indels of 1–16 bp in length. However, large indel events were not extensively studied by the authors. For NGS data, there are a number of ways to detect large and long variants. These include analyses and features incorporated in genome assembly (complete de novo or using unmapped reads only), read splitting, read coverage depth analysis, inconsistencies of insert sizes through paired-end mapping, etc. Here we will briefly introduce the major methodologies in the field as well as discuss their advantages and disadvantages.

Read-depth methods

In diploid or polyploid organisms, duplicated or deleted regions will show significantly higher or lower coverage. Therefore, one can count the number of short sequencing reads to estimate the relative ploidy or copy number of large genomic regions. In general read-depth approaches assume a random distribution in mapping depth and a per-sample GC correction of read-depth signal is typically performed to increase the signal–noise ratio. Read-depth methods are effective for large (>1 kb) copy number variants (CNVs) but are not able to identify copy number neutral variants like inversions. Its resolution is the poorest among all current approaches.

Read-pair methods

Read-pair methods are currently the most widely applied approach for identifying non-SNP variations. It uses mapping distance and the orientation of paired reads to statically assess the abnormity of mapping features. It is able to detect most kinds of structural variants. However it is not effective for very small variants or very large insertions. Also it could not precisely identify the breakpoints of reported variants.

Split-read methods

The presence of variants leaves a signal in short-read data that can be found with a split-read mapping. By breaking short reads into two or more fragments and aligning them separately on the reference genome, the precise location, size and types of variants can be computed efficiently. The split-read approach is powerful for detecting small and medium-size variants and is able to pinpoint the precise breakpoints. It is, however, less effective for detecting large variants and those in repetitive regions.

Assembly

De novo assembly from short reads perhaps offers the best way to detect long insertions and structural rearrangements. There are a number of short read assemblers based on de Bruijn graphs (Zerbino and Birney, 2008; Li *et al.*, 2010; Generre, *et al.*, 2011; Simpson and Durbin, 2012). Recently, Li *et al.* (2011) presented an analysis on two human genomes by de novo assembly with single base resolution, and the assembly algorithms developed by Iabal *et al.* (2012) enabled the detection of both simple and complex genetic variations in an individual or population. However, without the reference genome, Li and colleagues' de novo assembly method is limited to homozygous alleles only. A different assembly based variant calling method (Iabal *et al.*, 2012) using a reference genome missed a significant amount of true variants. Furthermore, the read-pair information could not be used by this method in the process. Given the current typical read length of 35–150 bp and low read coverage in population genomic studies using the NGS platforms, the lack of high-quality variant calling using the current de novo assembly programs will likely continue in the near future. Therefore, the need is eminent in the genomic community to develop read mapping related algorithms in order to reliably identify structural variants.

Next we introduce Pindel, a split-read method that uses read mapping for the detection of indels and SVs.

Pindel: a pattern growth method to identify precise breakpoints of indels and SVs

In 2008, the read length of NGS platforms was as short as 36 bp and single ended. Mapping unique variants to reference sequences as large as 3 billion bases, such as the human genome, was a challenging task. At that time, the read-depth approach was the method of choice for CNV discovery. Later, paired-end sequencing techniques were invented and virtually doubled the read length. This enabled approaches making full use of read mapping distance and orientation between reads in a pair. At that stage, state-of-the-art mapping programs, like MAQ (Li *et al.*, 2008a) and SOAP (Li *et al.*, 2008b), were still not able to identify indels. Indeed, a split-read approach with 36 bp reads was considered impossible for detecting non-SNP variants, because sequences with half the length of such reads would map well to far too many locations. To help mapping short reads to unique locations reliably, an anchored split-read mapping, the so-called Pindel method, was invented to identify the breakpoints of large deletions (1 bp to 10 kb) and medium size insertions

(1–20 bp). We used pattern growth, originally a sequential pattern mining approach, for efficient string matching.

Pattern growth for string matching

Pattern growth is an approach that can mine patterns at various frequencies from large databases (Pei *et al.*, 2004). In 2007, this algorithm was implemented for the discovery of protein motif sequences (Ye *et al.*, 2007). Pattern growth can detect the minimum and maximum unique substrings of a given pattern from a sequence. The inputs are a pattern and a reference sequence while the outputs consist of all unique substrings together with their locations. The unique substrings can be searched from either the leftmost or the rightmost base of the pattern. Here, we briefly demonstrate how to find the unique substrings starting from the leftmost position of a pattern. The procedure is also illustrated in Table 6.1.

Let a reference sequence R be *'ATCGCGATCTCTTAGC'* and pattern P be *'CGATCG'*. The variable i is the growth index starting from 1. Θ is the set of locations where SP_i, the substring of pattern starting from position 1 to i, appears in R. First, SP_1 is '*C*', and the locations of '*C*' in R, which is (3, 5, 9, 11, 16), are stored in Θ. In the second loop, the algorithm will search for SP_2 '*CG*' in the reference and update Θ to (4, 6). As the procedure continues, the minimum unique substring $SP_{min} = SP_i$ when SP_i is the first unique substring. Later on, if SP_i is unique and SP_{i+1} is not, then SP_i is the maximum substring SP_{max}. In our example, SP_{min} = '*CGA*' and SP_{max} = '*CGATC*'. The pattern growth algorithm is implemented in the high-resolution remapping stage of the pipeline.

Table 6.1 An example of pattern growth. The inputs are a pattern '*CGATCG*' and a reference '*ATCGCGATCTCTTAGC*'

i	Subpattern (SP_i)	Θ	Feature
1	C	3, 5, 9, 11, 16	
2	CG	4, 6	
3	CGA	7	SP_{min}
4	CGAT	8	
5	CGATC	9	SP_{max}
6	CGATCG	Φ	

General procedure of Pindel

In the Pindel program, the goal is to compute the precise breakpoints as well as the fragments inserted or deleted compared to the reference genome from paired-end reads. In the preprocessing step, SSAHA2 (Ning *et al.*, 2001) was used to map all the reads to the reference genome. Then the mapping results are screened, keeping only those paired reads such that only one end can be mapped. For each of those read pairs, the mapped end must be uniquely located in the genome with no mismatched bases while the other end cannot be mapped to anywhere in the genome under a given threshold alignment score ($s = 20$ for ~ 36 bp reads). For each of those pairs, the Pindel program uses the mapped end to determine the anchor point on the reference genome and the direction of the unmapped read. Knowing the anchor point, the direction to search for the unmapped read and the user defined maximum deletion size (Max_D_Size) parameter, a subregion in the reference genome can be located, where Pindel will break the unmapped read into two (deletion) or three (short insertion) fragments and map the two terminal fragments separately.

Detecting large deletions

When paired-end reads are aligned to the reference genome, for the majority of the reads, both ends can be mapped to the reference genome. However, a small portion might have only one end mapped to the reference genome. One of the possibilities is that the unmapped read mate spans the breakpoint of a deletion event in the test sample compared to the reference genome as shown in Fig. 6.1. Thus, those unmapped reads could contain information about precise breakpoints of deletion events. However, further analyses are needed to identify a proper position to split the read into two fragments, which can then be mapped back to the reference separately, to compute the exact positions of the breakpoints as well as the deleted fragment in the sample that was present in the reference. The more reads support the same incidence, the greater the confidence is about the deletion event in the test sample.

For short reads (36 bp), half of a read length might be too short to be mapped uniquely in a whole genome as large as humans'. However, the

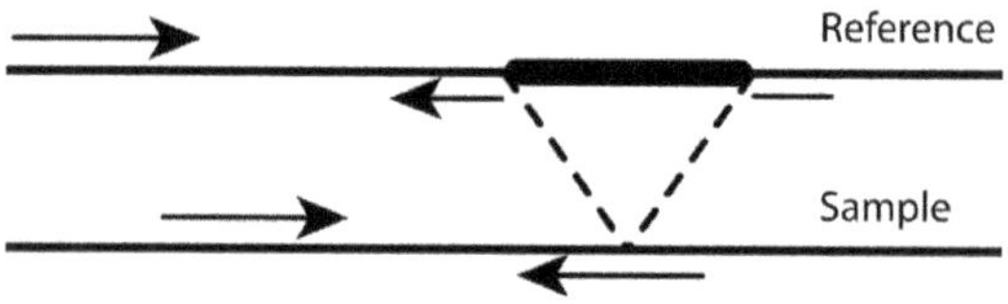

Figure 6.1 Detecting breakpoints of a deletion by Pindel. The unmapped read in the 'sample' genome carries the information about the precise breakpoints of the deletion event (from small to large numbers of bases). If we can find a proper position to split the read into two fragments and map them back to the reference separately, we will be able to compute the exact positions of the breakpoints as well as the deleted sequence in the 'sample' compared to the 'reference' genome.

location and the direction of the mapped read can reduce the search space dramatically. As shown in Fig. 6.1, when we have one read mapped (the anchor point) but the other one is not mappable, we only need to consider the local region at one side of the anchor point. The computational procedure for dealing with each unmapped read is described below:

1. Gather information about the location and direction of the mapped read from the mapping result obtained in the preprocessing step.
2. Define the 3′ end of the mapped read as the anchor point.
3. Use pattern growth algorithm to search for minimum and maximum unique substrings from the 3′ end of the unmapped read within the range of two times of the insert size from the anchor point.
4. Use pattern growth algorithm to search for minimum and maximum unique substrings from the 5′ end of the unmapped read within the range of read length + Max_D_Size starting from the already mapped 3′ end of the unmapped read obtained in step 3.
5. Check whether a complete unmapped read can be reconstructed by combining the unique substrings from the 5′ and 3′ ends found in steps 3 and 4. If yes, store the read and information in the database U. Note that exact matches and complete reconstruction of the unmapped read are required.
6. After processing all unmapped reads, sort database U according to the breakpoint coordinates in the reference and generate an output that includes every deletion event supported by at least two reads.

Detecting medium-sized insertions

For insertions in the medium size range (≥20 bases for 36 bp reads), the precise breakpoints and the size of the inserts can be computed. The computational procedure is very similar to that used for searching deletions. The only difference is in step 4 where the search range for the unique occurrence of minimum and maximum unique substrings from the 5′ end of the unmapped read is the read length minus 1. In this case, the complete read cannot be reconstructed and the extra bases belong to an inserted fragment compared with the reference genome as shown in Fig. 6.2.

The location and direction of the mapped read (green) define the local region to split the unmapped read. The 3′ end of the mapped read is defined as the anchor point. Then the pattern growth algorithm is used to search for minimum and maximum unique substrings from the 3′ end of unmapped reads within the range of two times of insert size starting from the anchor point. Using the pattern growth algorithm again, we can search for minimum and maximum unique substrings from the 5′ end of unmapped read within the range of read length −1, starting from the already mapped 3′ end of the unmapped read. The computed minimum and maximum substrings from both the 3′ and 5′ ends are examined to see whether they are adjacent to each other. The middle fragment is the inserted fragment. All possible solutions are stored in a database for

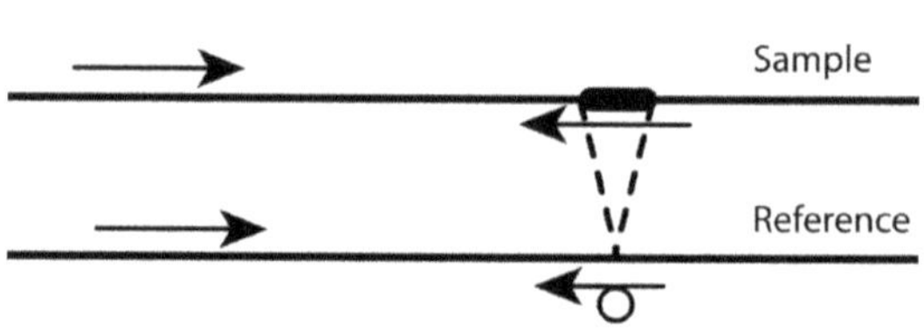

Figure 6.2 Detecting breakpoints of a short insertion by Pindel.

sorting according to the breakpoint coordinates. An insertion event is reported if at least two reads support it.

Performance on real data (NA18507)

The whole-genome sequence (Bentley *et al.*, 2008) of a male Yoruba from Ibadan, Nigeria (YRI, sample NA18507), provided an opportunity to examine the performance of Pindel using deep coverage, high-quality, and most importantly real paired-end read data. After preprocessing the paired-end reads with SSAHA2, 56,161,333 pairs of reads each had only one end mapped uniquely to the human reference genome while the other end could not be mapped. The search was constrained to short indels (1–16 bp) to allow direct comparison with results reported in the original article. Our analyses using Pindel predicted 146,843 deletions and 142,908 insertions. In total, 133,974 deletions (91.2% of 146,843) and 124,559 insertions (87.2% of 142,908) in our analyses overlapped those reported in Bentley *et al.* Thus, if the predictions reported by Bentley *et al.* were all true positives, the overall false negative rate of Pindel for deletions and insertions of length 1–16 bp is 31.6% for this dataset.

Since the first version of Pindel is able to identify deletions of sizes up to 10 kb from 36 bp paired-end reads, larger deletions were compared with the database of genomic variants, DGV (http://dgv.tcag.ca/dgv/app/home/). For deletions in the range of 100 bp-1 kb, Pindel predicted 1,399 deletions, 949 of which (67.8%) were listed in DGV. For those in the range of 1–10 kb, Pindel predicted 1138 deletions, 494 of which (43.4%) were listed in DGV. The deletion and insertion size distribution from 1 bp to 10,000 bp for NA18507 is shown in Fig. 6.3, where the frequency corresponding to each size is depicted as a dot while adjacent dots are connected to illustrate the trends. There is a peak around 300 bp, which contains hundreds of putative SINEs.

Recent developments

Besides the published methods for computing deletions and short insertions, recently we substantially improved the methodology for the detection of additional variant types. For example we developed a procedure to compute the precise

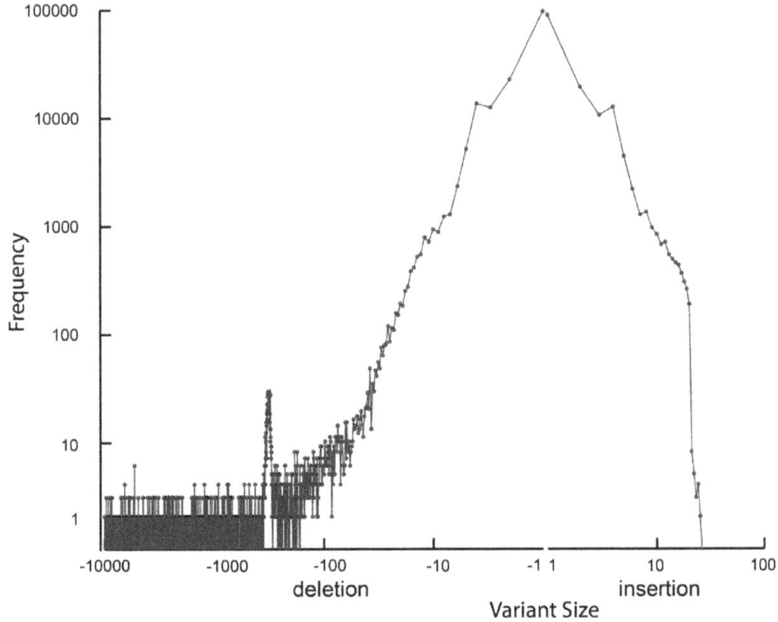

Figure 6.3 Plots of deletion size distribution for NA18507 from 1 to 10,000 bp and insertions from 1 to 25 bp. There is a peak around 300 bp deletion, which may contain hundreds of putative SINEs.

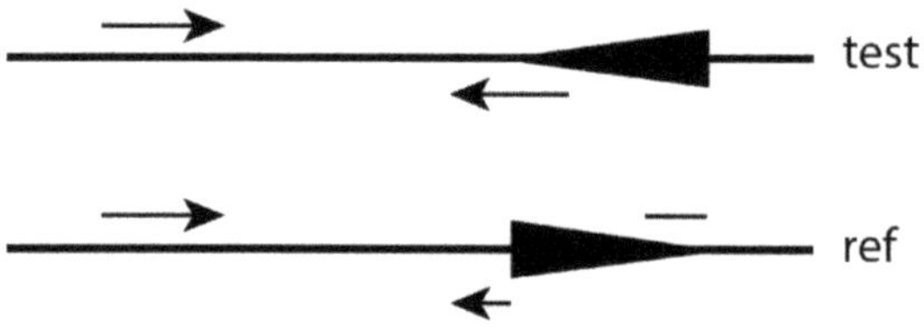

Figure 6.4 Detecting other types of variants with inversion as an example. For example, if the genome of a patient (test) has an inversion relative to the reference genome (ref), current mapping algorithms are unable to find that inversion as part of the read is on the 'wrong' strand and also reversed. Through Pindel, the second fragment would be detected as an inversion, as well as locating the precise breakpoints and the size of the inverted fragment.

breakpoints of inversions, as shown in Fig. 6.4. The fragments represented by triangles correspond to reverse complementary sequences in the test sample compared to the reference genome. By mapping the read across the breakpoint as two fragments on different strands of the genome, the precise breakpoints, size and sequence can be computed.

Further advances of split-read approaches

Through Pindel, it was demonstrated that the split-read approaches could be fruitful for identifying non-SNP variants, even using short NGS reads of paired-end 36 bp. However, the first version of Pindel did not tolerate sequence errors and was optimized only for specificity but not for sensitivity. To improve sequence-error tolerance and sensitivity, other methods have been developed to further explore the power of split-read methods. Below we briefly describe these methods.

SRiC

Zhang *et al.* (2011) developed a method called split-read identification calibrated (SRiC) for detecting short insertions and deletions. The first step using this method is to map the 35-mer of each end of a 454 read to a whole reference genome using Bowtie. Then BLAT is applied to obtain a gapped alignment. As 454 reads are much longer than those from other NGS platforms, the unique mapping of partial reads is less problematic while sequencing and mapping errors are modelled so as to ensure correct structural variant calls.

SLOPE

The SLOPE program, implemented in C++, processes unmapped reads for the identification of short indels and translocations (Abel *et al.*, 2010). This program allows both substitutions and gapped alignment for string matching. Perhaps this is the reason that it is less sensitive (~ 50%) to small indels (1–5 bp) compared to about 95% sensitivity for other size ranges. This program is optimized for targeted re-sequenced DNA and it can quickly and accurately define translocation breakpoints.

SVseq

Zhang and Wu (2011) applied the Burrow–Wheeler transform (BWT) algorithm to split-read mapping and developed a program called SVseq. The SVseq program can find exact breakpoints of deletions. In general, SVseq is faster than Pindel and generates a greater number of positive calls at the expense of a higher false-positive rate.

SplazerS

Both the counting q-gram matches (Burkhardt *et al.*, 1999) and the SWIFT filtering algorithm (Rasmussen *et al.*, 2005) are implemented in the SplazerS program. This program searches for split read matches where the prefix and suffix match of a read may be interrupted by a large gap representing a deletion. Here, the prefix and suffix matches are respectively identified with the q-gram filtering criteria and only regions satisfy both prefix and suffix filter matches are considered. Both mismatches and gaps are allowed during string matching and thus the sensitivity of SplazerS is improved over Pindel at the expense of runtime.

CREST

Wang *et al.* (2011) proposed a method called CREST to make use of clipping reads for the detection of structural variants at the nucleotide level. As modern short read alignment programs are able to handle small indels without providing a soft clipping signature, CREST is not sensitive for detecting small structural variants. However, it can effectively detect translocations.

Conclusion and future perspectives

Compared to read-depth and read-pair methods, split-read approaches are able to detect small indels and small to medium-sized structural variants at the nucleotide level. However, relying on partial mapping of short reads can potentially bring in false positives, especially in repetitive regions. On the other hand, read-depth and read-pair approaches in general examine variants globally and thus show less bias towards repetitive elements. Note that none of read-depth, read-pair, split-read and assembly approaches is comprehensive enough for identifying all the different types of structural variants. Recently several attempts have been reported to incorporate multiple techniques to improve the sensitivity and specificity of structural variant detection. For example, GenomeSTRiP (Handsaker *et al.*, 2011) combines read-pair and read-depth methods and performs well for detecting deletions. Further investigations on combining the above four approaches are needed in order to derive a comprehensive and accurate method.

DNA sequencing is a dynamic and fast moving area with technologies and platforms being updated at an unprecedented pace. Accordingly, algorithms for identifying complex variants need to be improved as well. Recently the Illumina platform enhanced the performance of MiSeq, which is capable of producing 2×250 bp reads and triples the throughput (see Chapter 1). Ion Torrent from Life Technologies can decode a human genome in one day at a cost of ~ $1000. As read length increases, the numbers of reads and read pairs can decline accordingly, if we were to maintain the same total coverage. Thus, split-read and assembly approaches will likely play a more important role for detecting all variant types. While the pattern growth algorithm in Pindel can handle substitution errors, it is less effective at dealing with indel errors, which are rather common in data generated by the Ion Torrent and PacBio systems. An effective and efficient string-matching algorithm is essential for optimal performance on these new types of sequencers. Several algorithms reviewed here (e.g. q-gram and BWT) have shown promise of capable of handling such sequencing errors.

As the throughput of sequencers increases and the price of sequencing genomes drops, more genome data will be generated at a higher rate. The storage and processing of genome sequence data remain the bottleneck. Before genomic data can be utilized for clinical applications, a faster and less memory-intensive genome analysis software package is needed to process hundreds of genomes within days using a modest computer cluster with about 100 CPUs. Investigations so far have demonstrated the potential of the split-read approach and we believe this approach could be the core of such a package.

References

Abel, H.J., Duncavage, E.J., Becker, N., Armstrong, J.R., Magrini, V., and Pfeifer, J.D. (2010). SLOPE: a quick and accurate method for locating non-SNP structural variation from targeted next-generation sequence data. Bioinformatics *26*, 2684–2688.

Bennett, E.A., Coleman, L.E., Tsui, C., Pittard, W.S., and Devine, S.E. (2004). Natural genetic variation caused by transposable elements in humans. Genetics *168*, 933–951.

Bentley, D.R., Balasubramanian, S., Swerdlow, H.P., Smith, G.P., Milton, J., Brown, C.G., Hall, K.P., Evers, D.J., Barnes, C.L., Bignell, H.R., *et al.* (2008). Accurate whole human genome sequencing using reversible terminator chemistry. Nature *456*, 53–59.

Burkhardt, S., Crauser, A., Ferragina, P., Lenhof, H., Rivals, E., and Vingron, M. (1999). Q-gram based database searching using a suffix array (Quasar). In RECOMB, 77–83.

Gnerre, S., MacCallum, I., Przybylski, D., Ribeiro, F.J., Burton, J.N., Walker, B.J., Sharpe, T., Hall, G., Shea, T.P., Sykes, S., *et al.* (2011). High-quality draft assemblies of mammalian genomes from massively parallel sequence data. Proc. Natl. Acad. Sci. U.S.A. *108*, 1513–1518.

Handsaker, R.E., Korn, J.M., Nemesh, J., and McCarroll, S.A. (2011). Discovery and genotyping of genome structural polymorphism by sequencing on a population scale. Nature Genet. *43*, 269–276.

Iafrate, A.J., Feuk, L., Rivera, M.N., Listewnik, M.L., Donahoe, P.K., Qi, Y., Scherer, S.W., and Lee, C. (2004). Detection of large-scale variation in the human genome. Nat. Genet. *36*, 949–951.

Iqbal, Z., Caccamo, M., Turner, I., Flicek, P., and MacVean, G. (2012). De novo assembly and genotyping of variants using colored de Bruijn graphs. Nature Genetics *44*, 226–232.

Kidd, J.M., Cooper, G.M., Donahue, W.F., Hayden, H.S., Sampas, N., Graves, T., Hansen, N., Teague, B., Alkan, C., Antonacci, F., *et al.* (2008). Mapping and sequencing of structural variation from eight human genomes. Nature *453*, 56–64.

Levy, S., Sutton, G., Ng, P.C., Feuk, L., Halpern, A.L., Walenz, B.P., Axelrod, N., Huang, J., Kirkness, E.F., Denisov, G., *et al.* (2007). The diploid genome sequence of an individual human. *PLoS Biol.* 5, e254, doi:07-PLBI-RA-125810.1371/journal.pbio.0050254.

Li, H., Ruan, J., and Durbin, R. (2008a). Mapping short DNA sequencing reads and calling variants using mapping quality scores. Genome Res. *18*, 1851–1858.

Li, R., Li, Y., Kristiansen, K., and Wang, J. (2008b). SOAP: short oligonucleotide alignment program. Bioinformatics *24*, 713–714.

Li, R., Zhu, H., Ruan, J., Qian, W., Fang, X., Shi, Z., Li, Y., Li, S., Shan, G., Kristiansen, K., *et al.* (2010). De novo assembly of human genomes with massively parallel short read sequencing. Genome Res. *20*, 265–272.

Li, Y., Zheng, H., Luo, R., Wu, H., Zhu, H., Li, R., Cao, H., Wu, B., Huang, S., Shao, H., *et al.* (2011). Structural variation in two human genomes mapped at single-nucleotide resolution by whole genome de novo assembly. Nature Biotecnol. *29*, 723–730.

Mills, R.E., Luttig, C.T., Larkins, C.E., Beauchamp, A., Tsui, C., Pittard, W.S., and Devine, S.E. (2006). An initial map of insertion and deletion (INDEL) variation in the human genome. Genome Res. *16*, 1182–1190.

Ning, Z., Cox, A.J., and Mullikin, J.C. (2001). SSAHA: a fast search method for large DNA databases. Genome Res. *11*, 1725–1729.

Rasmussen, K.R., Stoye, J., and Myers, E.W. (2006). Efficient q-gram filters for finding all spsilon-matches over a given length. J. Comput Biol. *13*, 296-308.

Simpson, J.T., and Durbin, R. (2012). Efficient de novo assembly of large genomes using compressed data structures. Genome Res. *22*, 549-56.

Wang, J., Mullighan, C.G., Easton, J., Roberts, S., Heatley, S.L., Ma, J., Rusch, M.C., Chen, K., Harris, C.C., Ding, L., *et al.* (2011). CREST maps somatic structural variation in cancer genomes with base-pair resolution. Nat. Methods *8*, 652–654.

Ye, K., Kosters, W.A., and Ijzerman, A.P. (2007). An efficient, versatile and scalable pattern growth approach to mine frequent patterns in unaligned protein sequences. Bioinformatics *23*, 687–693.

Ye, K., Schulz, M.H., Long, Q., Apweiler, R., and Ning, Z. (2009). Pindel: a pattern growth approach to detect breakpoints for large deletions and medium sized insertions from paired-end short reads. Bioinformatics *25*, 2865–2871.

Zerbino, D.R., and Birney, E. (2008). Velvet: algorithms for de novo short read assembly using de Bruijn graphs. Genome Res. *18*, 821–829.

Zhang, J., and Wu, Y. (2011). SVseq: an approach for detecting exact breakpoints of deletions with low-coverage sequence data. Bioinformatics *27*, 3228–3234.

Zhang, Z.D., Du, J., Lam, H., Abyzov, A., Urban, A.E., Snyder, M., and Gerstein, M. (2011). Identification of genomic indels and structural variations using split reads. BMC Genomics *12*, 375.

Novel Insights from Re-sequencing of Human Exomes Through NGS

7

Tao Jiang, Jun Li, Xu Yang and Jun Wang

Abstract

The advent of exome sequencing has opened a new door for utilizing next-generation sequencing (NGS) in human disease research. As an efficient and cost-effective method, exome sequencing based on NGS platforms has brought human disease research into a new era. Since many human diseases are caused by mutations in protein-coding regions, exome sequencing is regarded as a powerful approach to study human diseases. Several exome capture kits have been developed to facilitate exome sequencing. The wide application of exome sequencing in recent years has significantly boosted scientific discoveries. Through bioinformatic analyses after exome sequencing, many genes are implicated in human diseases. As NGS price decreases rapidly, exome sequencing is expected to become a routine method for academic and clinical use.

Introduction

It is well known that only a small fraction (approximately 1%) of the whole human genome is expressed and translated into proteins. These protein-coding regions are named coding exons, and together, they constitute the 'exome'. On the one hand, the cost of sequencing the whole genome by NGS is still very high. On the other hand, it is reported that most mutations causing human diseases are located in protein-coding exons. Thus, sequencing coding exons is very likely to detect disease-related genes, and this has led to the emergence of exome sequencing. Compared to whole genome sequencing, the cost of sequencing just the protein-coding portion of the entire genome by NGS is low enough for scientists to use exome sequencing as a routine tool in their researches. Accordingly, exome sequencing is an efficient and cost-effective way to reveal the pathogenesis of human diseases. Ever since its advent, the wide applications of exome sequencing in recent years have accelerated the discovery of disease-causing genes. These discoveries in various areas of human disease research will be described in the sections below.

A significant step in exome sequencing is to capture/enrich the sequences of coding exons in the entire genome for sequencing. Thus, an important requirement is the development of efficient exome sequence capture/enrichment tool kit. At present, most kits came from three companies. In spite of small differences between them, the fundamental principle is more or less the same. Specifically, the sample DNA is first turned into single-stranded genomic DNA. The single-stranded DNA molecules are hybridized with probes that are complementary to the sequences of all coding exons in the genome. Then, the hybridized mixture is magnetically separated to obtain the enriched exons, which is subsequently sequenced by NGS instruments. In addition to the set of known coding exons (~30 Mb), these kits could capture a few non-coding regions, e.g. promoters and untranslated regions (UTRs).

Our chapter on exome sequencing is divided into the following six sections. The protocol section describes the steps and principles of exome sequencing. The second section introduces the main exome capture platforms and commercial kits in the market. The third section provides an overview of quality control and performance evaluation. The fourth section describes the bioinformatics analysis pipeline for exome sequencing. The fifth section reviews the recent discoveries in several areas of human diseases research revealed by applying exome sequencing. The last section provides a perspective of exome sequencing.

The protocol

This section will illustrate the processing of DNA samples for exome capture followed by next-generation sequencing. The first step to conduct exome sequencing research is to enrich the targeted exome region in the whole genome. Before the advent of NGS, the enrichment is conventionally achieved by PCR amplification of target region. However, this method has the disadvantages of high cost and long operation time. It is not suitable for large-scale studies of human diseases (Mamanova *et al.*, 2010).

With the advent of NGS technologies, exome capture kits for massively parallel sequencing have been developed to accelerate disease researches. Qualified samples are processed for exome capture, library preparation and cluster generation. For each DNA sample to be sequenced, one library will be prepared. If the DNA sample contained RNA molecules (as revealed by the sample quality control screening), it would be first treated with RNase A, and then followed with sample preparation procedure. The basic principle of exome sequencing is similar among different exome capture platforms, as illustrated in Fig. 7.1. However, the equipment and reagents may differ among the platforms. The detailed description of the whole workflow using the Illumina NGS platform and Agilent SureSelect Human All Exon Kit will be shown below to demonstrate the protocol. We would like to mention that we are describing the workflow but not a step-by-step protocol, which can be found in the manuals of commercial exome enrichment kits.

DNA fragmentation

1 DNA will be fragmented to 150–200 bp by focused ultrasound or divergent ultrasound using a Bioruptor or Covaris S2/E210 shearing device. Other matching fragmentation

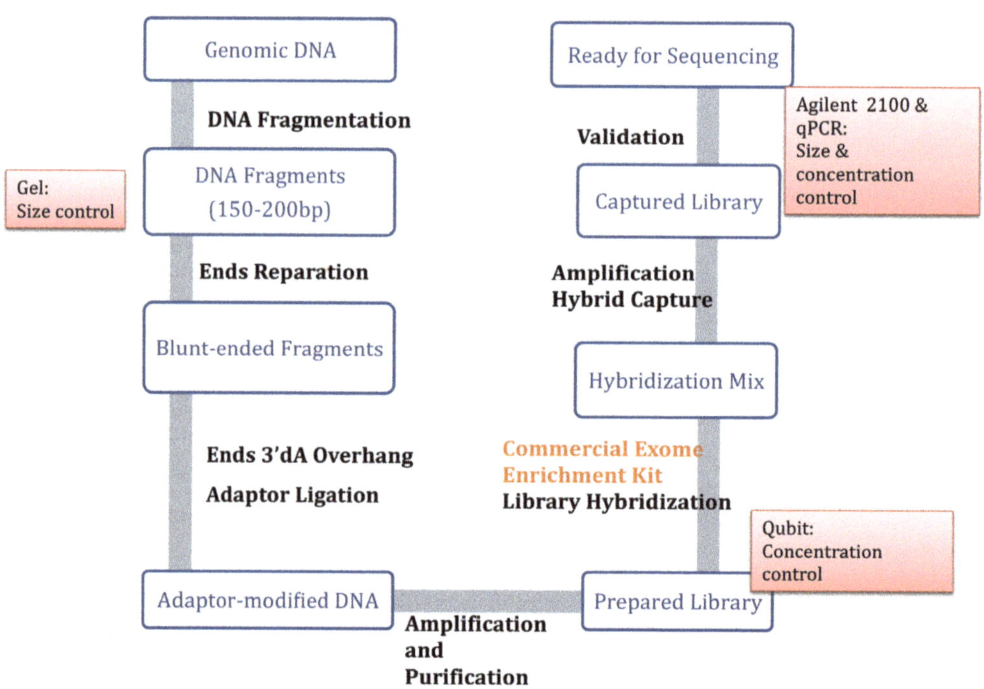

Figure 7.1 The workflow and quality control points during sample preparation in exome sequencing. The red frames indicate the points of quality control.

systems can also be used and appropriate parameters will be set to control the size range of fragmented DNA.

2 The sizes of fragmented DNA molecules will be checked on a gel. If the average size is not as expected, the sheared product will be re-fragmented (when they are too big) or the shearing parameters will be modified (when the sheared fragments are too small) to work on the original sample.

Ends repair

1 The repairing reaction system will be prepared and incubated at 20°C for 30 minutes in a Thermomixer. The key enzymes include T4 DNA Polymerase, Klenow Fragment and T4 Polynucleotide Kinase. Then, the blunt ends are generated.

2 The blunt-ended DNA products will then be purified with a purification kit.

Ends with the 3′-dA overhang

1 An 'A' base will be added to the 3′ ends of the blunt phosphorylated DNA fragments using a DNA polymerase. This step promotes the DNA fragments to be ligated with the adapters, which have a single 'T' base overhanging at the 3′ end.

Main reagents include: 10x blue buffer, dATP and the Klenow fragment.

2 Products after this step will need to be purified using a purification kit.

Adapter ligation

1 The ligation reaction system will be prepared and incubated at 20°C for 15 minutes in a Thermomixer to ligate adapters to the ends of the DNA fragments. Based on the type of library that you are planning to prepare, you can choose either the Illumina PE or Index PE adapters.

2 The products of the ligation reaction will be purified with a purification kit.

Amplification and purification

1 DNA fragments with adapters on both ends will be amplified by PCR using the Phusion DNA Polymerase. Time for the elongation step of the PCR depends on the average size of DNA fragments in the library.

2 To prepare index libraries, index tags are introduced into the adapters at this stage by primers containing appropriate barcode tags.

3 The PCR products are then purified with a purification kit and the concentration of the purified PCR products will be controlled with Qubit. The library for exome capture is ready for the next step.

Library hybridization

1 The prepared DNA library is mixed with the capture probe and hybridization buffer blocks in a separate PCR plate according to the SureSelect Target Enrichment System for Illumina Paired-End Sequencing Library Protocol.

2 The hybridization is performed on a thermal cycler for 24 hours at 65°C using the Agilent SureSelect All exon kit.

Hybrid capture

1 The probe-targeted DNA from the hybridization-mixture is separated using Dynal MyOne Streptavidin T1 (Invitrogen).

2 The probe-targeted DNA with the Dynal beads is released through washing by elution buffer.

Amplification of the captured library

1 The captured library is amplified with PCR using the Herculase II Fusion DNA Polymerase.

2 Then, the captured library is purified with the Agencourt AMPure beads. And the captured library is now completely prepared.

Validation and sequencing of the library

1 The size and concentration of DNA in the captured library are determined by Agilent 2100 Bioanalyzer and ABI StepOnePlus Real-Time PCR System.
2 Non-barcoded libraries can proceed to flow cell preparation and index-tagged libraries are pooled at different ratios based on the expected amount of data to be generated in sequencing.

Equipment

See Table 7.1.

Exome capture platforms and kits

Solid-based vs. solution-based platforms

For exome capture platforms, the method for enrichment is very important, because they can affect the efficiency of exome capture. Until now, both solid and solution-based methods have been used for exome enrichment. Because NGS platforms are continuously upgraded, each company typically has several versions of exome capture platforms as well. The solid-based methods were developed earlier than solution-based methods for exome enrichment, with NimbleGen 2.1M Human Exome Arrays as the first commercialized whole exome capture product (http://www.nimblegen.com/products/seqcap/index.html). In the solid-based methods, genomic DNA is fragmented and hybridized with probes in a solid-phase exome array and the target exome region is captured by hybridization. Unbound DNA fragments are removed by washing. Then, the enriched DNA fragments pool is eluted and subsequently amplified by PCR for library construction. After quality control checking, the library is ready for sequencing by NGS platforms (Hodges *et al.*, 2007).

However, due to recent developments in NGS technology, solid-phase exome arrays are becoming increasingly insufficient for high-throughput NGS researches. Thus, an alternative system of solution-based capture method has been developed for exome enrichment, such as the Agilent SureSelect Target enrichment system (http://www.halogenomics.com/sureselect/how-it-works). In the solution-based method, probes and beads are placed in solution in an Eppendorf tube and the target exome region is enriched by hybridization with probes. The complexes of probes and enriched DNA fragments are pulled down by beads, and unbound DNA fragments are removed by washing. Then, the enriched DNA fragments pool is eluted. The follow-up steps of library construction and sequencing are similar to those in solid-based methods (Gnirke *et al.*, 2009). At present, solid-based methods are rarely used while solution-based methods are increasingly used. Compared to solid-based methods, the solution-based systems exhibit the following advantages:

- Easy to operate and automate;
- No specific system and equipment for hybridization;
- High throughput, not limited by equipment;
- Less amount of DNA sample; and
- Short hybridization period.

Table 7.1 Required equipment for library preparation in exome sequencing

Equipment	Vendor and part number
Agilent 2100 Bioanalyzer	Agilent p/n G2938C
Thermal cycler	Applied Biosystems Veriti Thermal Cycler
Covaris S-series Single Tube Sample Preparation System, Model S2	Covaris
Covaris microTUBE with AFA fibre and snap cap	Covaris p/n 520045
Microcentrifuge	Eppendorf Microcentrifuge Model 5424
Qubit Fluorometer	Invitrogen p/n Q32857
Dynal DynaMag-2 magnetic stand	Invitrogen p/n 123–21D or equivalent
Concentrator Plus	Eppendorf
Heat block	labnet

NimbleGen, Agilent, Truseq

There are several companies offering commercial exome enrichment kits, such as Agilent, NimbleGen and Illumina. And the platforms from Agilent and NimbleGen are widely used. The common products of NimbleGen and Agilent are introduced here based on their time of release.

NimbleGen

NimbleGen Sequence Capture Human Exome 2.1M Array

In 2008, NimbleGen Sequence Capture Human Exome 2.1M Array was developed to enrich target regions by hybridization with a solid-phase array. The kit could be customized to capture the target regions by hybridization with probes in arrays, exhibiting the characteristics of high specificity and coverage. This method saved time and effort in comparison with previous enrichment methods based on PCR. Since then, thousands of exome sequencing researches have been done with NimbleGen Human Exome 2.1M Array. The wide application of NimbleGen Human Exome 2.1M Array opened a new chapter in NGS-based human disease research, although it is rarely used now due to the limitations of solid-based methods.

NimbleGen SeqCap EZ Human Exome Library

In 2009, NimbleGen upgraded the exome capture platform to a solution-based method. The NimbleGen SeqCap EZ Library enables enrichment of the whole exome or customer regions of interest in a single test tube, such as NimbleGen SeqCap exome 44M V2. Built on an optimized design algorithm, the SeqCap EZ Library sets a new standard for a simple single-step enrichment method. The latest version is SeqCap EZ Human Exome Library v3.0. The new version offers a comprehensive coverage of coding regions with genes from the RefSeq RefGene CDS, CCDS, and miRBase v14 databases plus coverage of 97% Vega, 97% Gencode, and 99% Ensembl databases (http://www.nimblegen.com/products/seqcap/ez/v3/index.html).

Agilent SureSelect

The Agilent SureSelect target-enrichment kit is a solution-based exome capture platform, consisting of a series of exome kits, e.g. Agilent 38M, 44M and 50M. These kits could cover all coding exons or certain customized regions. These kits are unique in that they involve a RNA-driven DNA capture process. According to the protocol, each target enrichment kit is packed with a mixture of custom RNA oligonucleotides or 'baits' that are biotinylated for easy capture onto streptavidin-labelled magnetic beads. To capture the exome, genomic DNA is sheared and assembled into a library format specific to the sequencing instrument utilized downstream. RNA bait–DNA hybrids are then 'fished' out of the complex mixture by incubation with streptavidin-labelled magnetic beads and captured onto a strong magnet. After washing, the RNA baits on the beads are then digested so that the only remaining nucleotides are the targeted DNA of interest. With recent technical developments, the new target enrichment kits can cover even larger regions. For example, SureSelect Human All Exon V5 can cover a comprehensive 71 Mb, consisting of exons and UTRs (http://www.halogenomics.com/sureselect/all-exon-kits).

TruSeq

Illumina offers the TruSeq exome enrichment kit, an in-solution sequence capture platform for isolating exonic regions of interest in the human genome using hybrid selection. Prior to exome enrichment using this system, libraries are prepared using the TruSeq DNA sample preparation kit. The kit provides a robust indexing solution for each DNA sample, enabling multi-sample pooling of up to six samples in a single enrichment reaction. The ability of multi-sample pooling dramatically reduces hands-on time compared to other available methods, increasing the throughput and decreasing the cost of sequencing studies. After the DNA sample preparation step, targeted regions are captured with TruSeq exome enrichment kit. The enriched DNA fragments are amplified for cluster generation and subsequent sequencing. The DNA input is similar to that for other kits. With TruSeq DNA sample preparation kit, it can even start from 1 μg of original DNA (http://www.illumina.com/truseq.ilmn).

The main characteristics of several solution-based exome capture platforms are presented in Table 7.2 for comparison (Table 7.2).

The probes in exome enrichment kits are used to capture exome regions in the genomes of interest. The kits are designed using one or more of the following databases as references:

CCDS (Consensus CDS): a core set of human and mouse protein-coding regions that are consistently annotated and of high quality, consisting of over 18,606 human and over 19,988 mouse genes;

miRBase (Sanger): an comprehensive database of published miRNAs, including sequences, annotations, and target gene predictions. miRBase v20 was released in June 2013, containing 24,521 entries representing hairpin precursor miRNAs and expressing 30,424 mature miRNA products in 206 species.

RefSeq: an integrated, non-redundant and well-annotated database of naturally occurring DNA, RNA, and protein sequences. It was built by NCBI to provide only one representative reference sequence of each gene to avoid duplication for major organisms ranging from viruses to bacteria to eukaryotes.

GENCODE (Sanger): This database includes all evidence-based gene features in the entire human genome at a high accuracy, including all protein-coding loci with alternatively transcribed variants, non-coding loci with transcript evidence, and pseudogenes.

Table 7.2 General information of common exome-capture platforms

Exome in-solution capture platforms	Agilent SureSelect			NimbleGen SeqCap	Illumina TruSeq
	38M V1	44M V2	50M	44M V2	62M
Databases targeted	CCDS Sep 2008, miRBase v13	CCDS Sep 2009, RefSeq Jan 2010, miRBase v14	CCDS Sep 2009, GENCODE (including novel coding sequences validated by Sanger Institute), miRBase v13	CCDS Sep 2009, RefSeq Jan 2010, miRBase v14	CCDS Sep 2009, GENCODE, RefSeq Jan 2010, miRBase v13
Ref. genome[a]	Hg 18	Hg 19	Hg 18	Hg 19	Hg 19
Probe		RNA		DNA	DNA
Bait length		~120bp		60~90bp	95bp
Insert size[b]		150~200bp		200~300bp	~230bp
Library preparation		5 days		5 days	7 days[c]
Concentration		≥50ng/μl		≥50ng/μl	≥50ng/μl
Sample input		3μg		3μg	3μg[d]
No. of pooled samples within a single kit[e]		≥1		≥1	6

[a]The version of reference genome used by probe design.
[b]The DNA fragments length in DNA library constructed, which is closely related to the sequencing read length.
[c]Using TruSeq 62M kit, each round of library construction takes 7 days, because the samples need to be hybridized twice with the capture kit. Therefore, it takes more time than the other two platforms.
[d]Although Illumina claims that the DNA input of single TruSeq library construction starts from 1 μg, sometimes it is not feasible with this sample input amount. So 3μg DNA is still required for stable single library preparation.
[e]Number of samples participating in the same enrichment reaction. TruSeq enables multi-sample pooling with six samples (already barcoded) to hybridize with the same kit, which is the main reason of its low price. For NimbleGen and Agilent exome enrichment kits, two samples pooled hybridization is relatively stable. Pooled samples within a single kit are recommendable for collaborators who have large number of samples, but limited budget.

Quality control and performance evaluation

Quality control metrics

Several criteria are commonly used to measure the data quality of exome capture kits. These include:

- Specificity: the percentage of uniquely aligned reads that map to the designated probe region, relative to all reads uniquely aligned to the reference human genome.
- Uniformity: the relative abundance of reads distributed in the designed probe region, which could be intuitively evaluated by the distribution of per-base sequencing depth. But it is often measured by the distribution of cumulative depth (e.g., the percentage of target covered at least 10X).
- Sensitivity: the percentage of high-confident genotypes detected in all targeted bases. It mainly depends on depth and uniformity. The higher depth the capture data has and/or the more uniform the data is, the higher the sensitivity it would have.
- Accuracy: the concordance of detected SNPs from exome data in comparison with the genotyping array data.

While the four metrics listed above are the most critical, there are other general criteria including reproducibility, GC and allele biases. Recently, we comprehensively evaluated the relative performance for three main whole-exome capture platforms using the Illumina HiSeq2000 instrument: NimbleGen 2.1M Sequence Capture Array, NimbleGen SeqCap EZv1.0 and Agilent SureSelect 38Mb Kit. We found that the Agilent platform achieved higher capture specificity of targeted regions, but the two NimbleGen platforms showed better uniformity of coverage and greater genotype sensitivity, mainly due to their relatively high bait density. We also found that the three platforms had similar accuracy of genotype assignment and SNP detection. In addition, all three platforms showed similar levels of reproducibility, GC bias and reference allele bias (Asan *et al.*, 2011).

Here we extend the performance comparison to the updated exome capture platforms. All statistics are based on effective mean depths of 30X and 50X coverage. The 30X coverage here means that each base in the targeted exome is covered by 30 reads on average. Fig. 7.2 shows the comparison of four exome capture tools in several features including duplication rate, mapping rate,

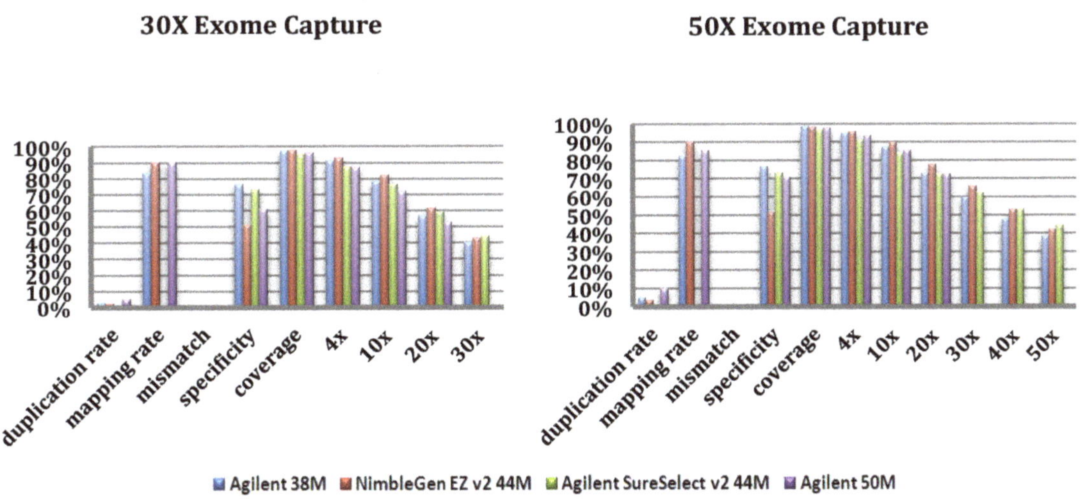

Figure 7.2 The comparison of four exome capture tools in terms of general quality control metrics at 30X (left) and 50X (right). The 4X shown in the x-axis indicates the fraction of target covered at least 4X, so is the same meaning for 10X, 20X, etc. The data presented here for each indicator is the mean value across 100 normal samples. Note that the data of duplication rate and mapping rate for the Agilent SureSelect v2 44M kit was missing, and the mismatch rate for all the platforms is less than 0.6% and therefore is not shown.

mismatch rate, specificity, coverage and cumulative depth distribution in target region. As can be seen from the figure, there are no significant changes among the capture tools for the first six indicators (duplication rate, mapping rate, mismatch rate, specificity, coverage, coverage ≥ 4x) with the increase of effective sequencing depth from 30X to 50X, indicating that sequencing depth has little influence on these indicators. However, the indicators at or above the cumulative depth of 10X, such as coverage ≥ 20X and coverage ≥ 30X, benefit more with the increase of depth coverage. As a result, higher depth will help to achieve higher fraction of exome covered at least 10X (~90% of target ≥ 10X for 50X mean depth using NimbleGen kit) and therefore high accuracy for detection of variants, especially for heterozygous SNPs. In addition, NimbleGen 44M kit shows the best performance except for specificity while Agilent 38M kit has the highest specificity among the four platforms, largely because the NimbleGen kit could capture much more off-target bases (Asan *et al.*, 2011). However, the specificity of the newest NimbleGen EZ capture kit (SeqCap EZ Exome v3.0) can be greatly improved after the optimization of experimental procedures, with much less sequencing effort to obtain similar fraction of target covered at least 10X compared with Agilent SureSelect 50M kit (http://www.nimblegen.com/exomev3launcheq/).

Sequencing depth selection

Both the coverage and cost will increase with sequencing depth. Therefore, it is necessary to determine what sequencing depth is enough to capture the exome region for subsequent researches. Through simulation and data sampling using the NimbleGen exome capture data, Fig. 7.3 below shows that the coverage in target region can reach 90% at a mean depth of 6X. When the average sequencing depth increases to 20X, the coverage of exome region by sequencing reads becomes almost saturated and reaches the practical upper limit (Jiang *et al.*, 2011). However, the increased depth helps to improve the sensitivity and accuracy of SNPs detection and identify rare disease-causing alleles. Therefore, 30X coverage is typically recommended for routine exome sequencing. Although Agilent exome kits have not been simulated, the relationship between coverage and mean depth is likely similar to that of the NimbleGen kit.

We also investigated the cumulative depth distribution under different mean sequencing depth using data from the research of 50 human exomes that revealed adaptation to high altitude (Yi *et al.*, 2010). The simulating graph confirms the similar conclusion that the fraction of exome region with at least 20X coverage increases slowly with average sequencing depth when the effective depth is greater than 100X (Fig. 7.4). The result suggests that there is also the practical upper limit for the cumulative depth distribution. Recently,

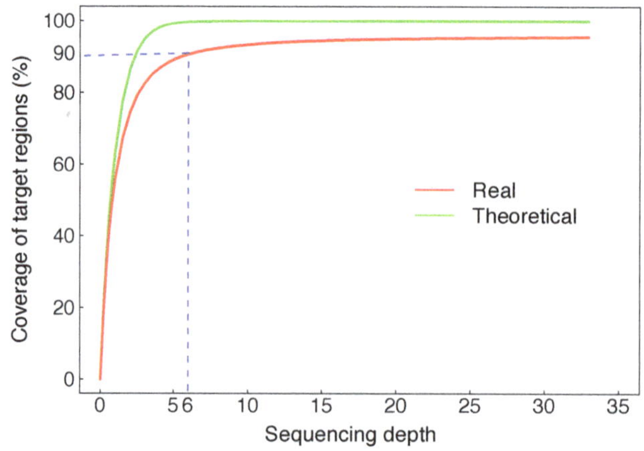

Figure 7.3 The theoretical and observed curves for the relationship between sequencing depth and coverage of exome.

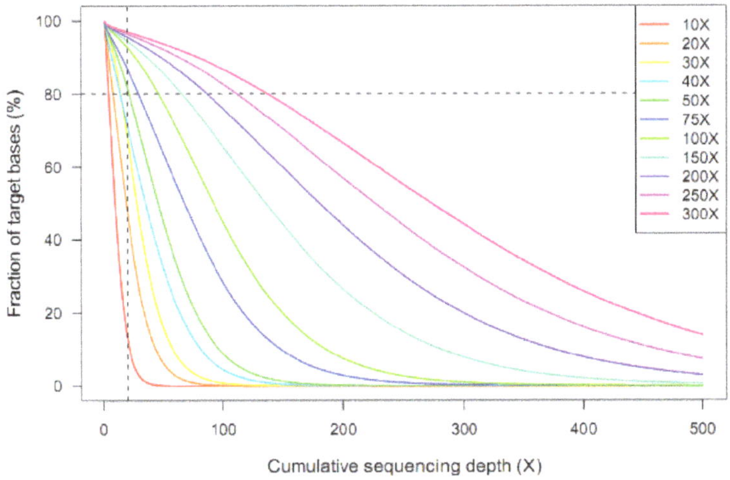

Figure 7.4 The cumulative sequencing depth distribution in the context of different mean sequencing depth. Y-axis indicates the fraction of target bases that achieves at or above a given sequencing depth. The vertical dash line indicates the cumulative depth of 20X, the horizontal dash line denotes the fraction of target covered by a specified cumulative depth is 80%. The intersection of the two lines means that 80% of the exome region will be covered by at least 20 reads when the exome is sequenced to a mean depth of 50-fold.

rare mutations have been revealed as closely related with the pathogenesis of human disease than previously expected. In order to identify rare disease-causing mutations, higher sequencing depth with NimbleGen or Agilent exome kits has been used in more and more research projects. From the papers published recently, 50X or higher sequencing depth (70–80X) was used in human disease research with median-scale samples by exome sequencing (Jones *et al.*, 2010). As the sequencing price decreases rapidly, it is possible to set 100X as the standard sequencing depth in the near future, especially for researches with small-scale samples.

Summary of exome capture kits

The Agilent SureSelect target enrichment kit is a commercially very successful exome capture system and it is widely used by the genomics research community. However, NimbleGen and TruSeq have features that make them appropriate for certain purposes (Table 7.3). Interestingly, the

Table 7.3 The advantages and disadvantages of Agilent, NimbleGen and Truseq exome capture kits

	Agilent 38M	Agilent 44M	Agilent 50M	NimbleGen 44M	TruSeq 62M
Advantage	High capture specificity, good quality control	Relatively more detected SNPs (a bit more than NimbleGen 44M)	Good coverage of Ensembl and RefSeq, novel coding exons included	The best comprehensive performance, good stability and uniformity	Cheap (6 pooled sample), large coverage of target region, low starting dose
Disadvantage	Low coverage, poor stability	Poor stability and uniformity	High duplication rate, bad uniformity, relatively few detected SNPs	Relatively low capture specificity of target region	Non-uniform pooling, twice hybridization, long period of library construction, high false positive
Customer	Limited budget but strict in false positive	Find as many SNPs as possible, prefer Agilent product	Abundant cost, prefer Ensembl and RefSeq	Prefer CCDS and RefSeq, care stability and uniformity	Limited budget and large sample size

capture systems are not limited to only coding exons but have expanded to non-coding region. For example, new exome capture kits, such as NimbleGen 64M and Agilent V4 (51M) & V5 (71M), can capture larger genomic regions, including non-coding segments. Depending on the purposes and research questions, different exome capture kits may be used.

Bioinformatics analysis

The conceptual bioinformatic analysis approach

Exome sequencing of a DNA sample will generate numerous variants from hundreds of millions of short DNA fragments in parallel. Most variants are already known as polymorphisms in human populations. Thus, the key challenge is to identify the disease-related variants among the background of non-pathogenic variants and sequencing errors (Bamshad *et al.*, 2011). The general conceptual approach for exome data analysis is illustrated in Fig. 7.5. After collecting the quality filtered reads from NGS instruments, diverse alignment softwares can be used to align the short reads to the human reference genome, the most popular aligners are listed in Table 7.4. Alignments are usually deposited in SOAP and SAM format, but then converted to sorted, indexed Binary Sequence Alignment/Mapping format (BAM) (Li *et al.*, 2009a) in order to save computational storage. The picard tool (http://sourceforge.net/projects/picard/) is then used to remove invalid alignments and remove duplicate reads from the BAM files before subsequent variants are identified. Based on the alignments, a series of SNP and Indel callers could be used to either identify high-quality SNPs and indels from each individual exome data or do SNP calling from multiple exome sequencing samples (Table 7.4). The final variants calls are represented in GFF format or standard Variant Call Format (VCF) (Danecek *et al.*, 2011).

The procedure of step-by-step analysis for exome data is complicated and time-consuming. To address this issue, some integrated pipelines based on the concept of parallel and cloud computing have been proposed, these include Crossbow (Langmead *et al.*, 2009), Atlas2 (Challis *et al.*, 2012) and HugeSeq (Lam *et al.*, 2012). The common SOAP package has already incorporated alignment, duplicates removal and variants discovery and filtering in one pipeline.

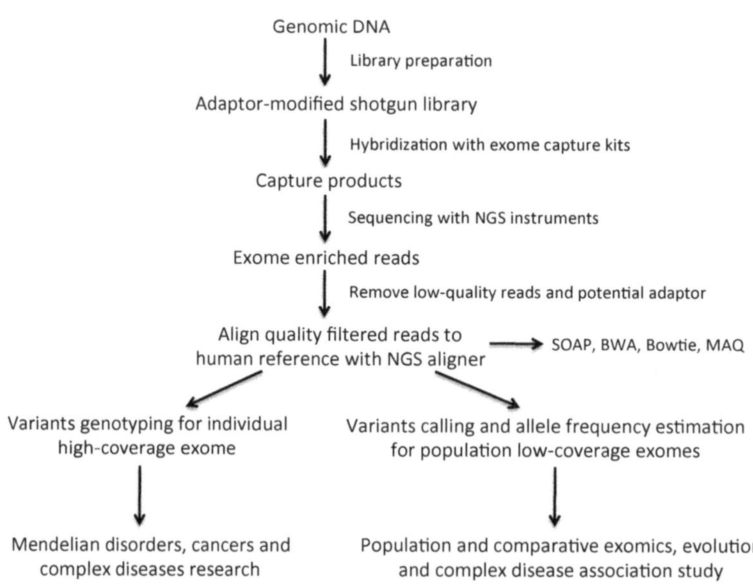

Figure 7.5 Overview of exome data analyses.

Table 7.4 Algorithms and software for exome data analyses

	BWT index	Hash based
NGS aligner	Bowtie, BWA, SOAP2	MAQ, ELAND, BFAST, Mosaik, mrFAST, Novoalign, Stampy
SNP caller	GATK, SOAPsnp, SAMTools, SNVer, GNUMAP-SNP	
Indel caller	GATK, Dindel, SAMTools, SOAPindel	
Multi-sample SNP caller and genotyping	SAMTools, GATK, Beagle, IMPUTE2, MaCH/Thunder	
Integrative pipelines	Crossbow, SOAP, Atlas2, HugeSeq	
Variants annotation and functional prediction	ANNOVAR, SIFT, PolyPhen2, GERP, PhyloP, PhastCons, VAAST	

Mendelian disease analysis as a demo case

The exome has been widely recognized as a source of rare disease-related variants and represents a highly enriched subset of the genome in which to search for variants with large effects. Thus, exome sequencing has been widely adopted in identifying causal genes responsible for Mendelian disorders (Bamshad *et al.*, 2011). Fig. 7.6 illustrates the general framework of exome sequencing in gene discovery responsible for monogenic traits.

As shown in Fig. 7.6, about 18,000~23,000 coding variants are commonly discovered per one exome by NGS tools. The variance in the number could be attributed to ethnic differences, with that African exome having the highest number of variants. Under the assumption that candidate variants involved in Mendelian disorders usually disrupt protein-coding sequences or normal splicing sites, synonymous variants would be far less likely to be pathogenic and thus be filtered in the subsequent analysis. Only non-synonymous variants (NS), splice acceptor and donor site mutations (SS), and coding indels (I) are considered. Next, the functionally significant variant set is further filtered against publicly available databases, such as dbSNP (Sherry *et al.*, 2001) and 1000 Genomes Project variants databases (Durbin *et al.*, 2010), as well as internal normal ethnic matched controls if possible. This step reduces the candidate pathogenic variants to around 300~500. With these novel rare variants in hand, researchers usually use some or several complementary functional prediction algorithms to predict the functionality of novel variants, either with approaches exploring the predicted changes in proteins caused by specific amino acid substitutions, such as SIFT (Ng and Henikoff, 2003) and PolyPhen2 (Adzhubei *et al.*, 2010), or by approaches that quantify nucleotide conservation under evolution, including PhastCons (Blanchette *et al.*, 2004), PhyloP (Pollard *et al.*, 2010) and GERP score (Cooper *et al.*, 2010). Some new integrated variant prioritization pipelines have been developed, such as ANNOVAR (Wang, K., *et al.*, 2010; Lyon *et al.*, 2011) and VAAST (Rope *et al.*, 2011; Yandell *et al.*, 2011). Through these analyses, rare deleterious candidate variants are further reduced to 200~250. Since there are differences in the modes of inheritance between a trait and the causal variant of either a *de novo* or inherited variants, the candidate variants will be analysed by different strategies according to disease trait. These strategies include using unrelated cases, family cases, and parent–child trios. After this step, the compelling candidate variants or genes need to be validated in additional larger pedigree and/or sporadic cases and controls to further demonstrate its causality for disease trait. In the end, in order to further investigate the biological role of the putative causal gene, *in vivo* functional study should be conducted to reveal the disease mechanism.

Here we describe how to use the framework discussed above to identify the pathogenic mutation in the *MVK* gene responsible for disseminated superficial actinic porokeratosis (DSAP) (Zhang *et al.*, 2012a). Using the Agilent SureSelect Human All Exon Kit (in solution), we subjected three individuals in the focused pedigree for exome capture, two of which suffered from DSAP, an autosomal dominantly inherited epidermal keratinization disorder. Exome sequencing was performed to achieve at least 50-fold coverage on average for each

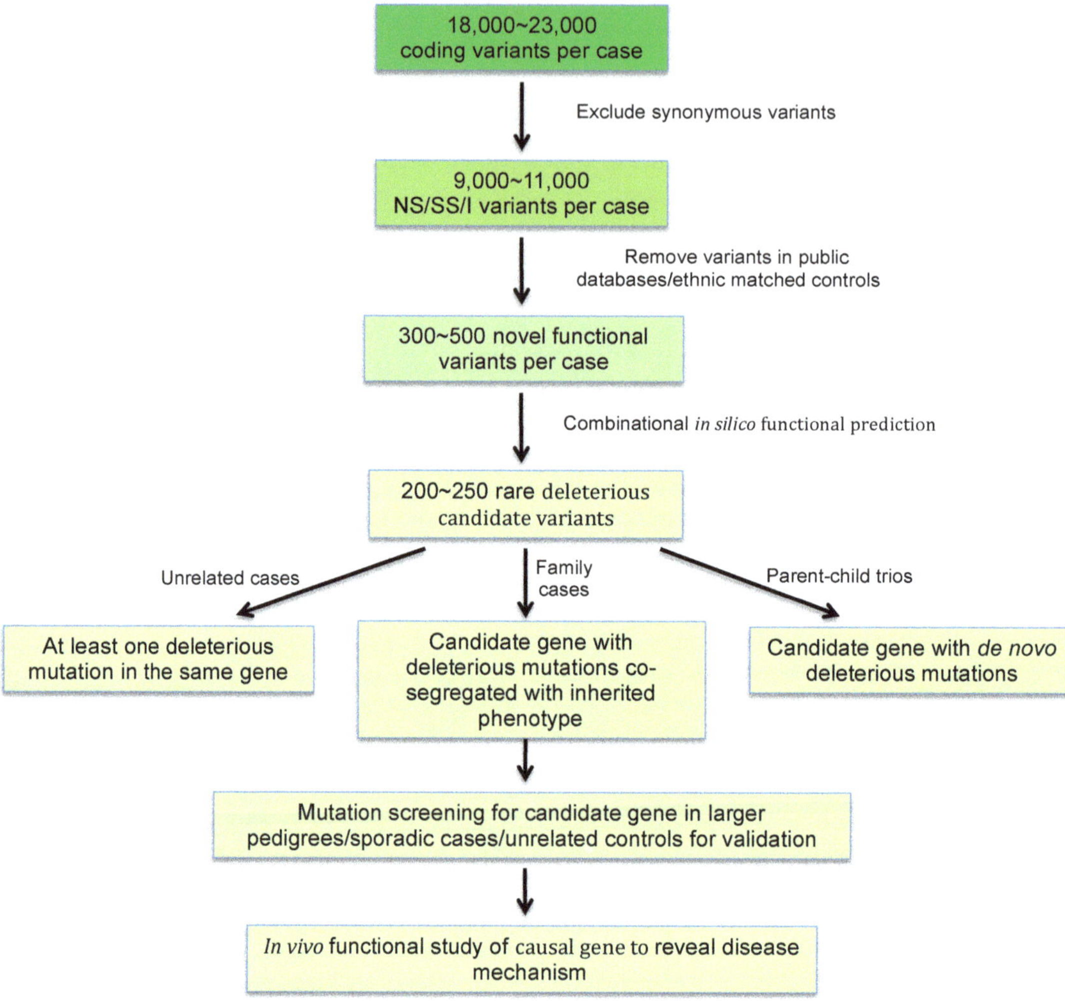

Figure 7.6 The general analytical framework for identifying genes responsible for monogenic (i.e. Mendelian) diseases through exome sequencing.

participant. Through variant detection, filtering and functional predictions, we identified a novel deleterious mutation (c.764T>C) in gene *MVK*. This gene was located in the previously identified linkage genomic regions and therefore considered as the candidate gene for this disorder in our examined family (Table 7.5). The mutation in *MVK* was first tested for cosegregation with the DSAP phenotype in the extended pedigree of this family by Sanger sequencing. All six affected individuals were heterozygous for this mutation, and it was absent in the three unaffected family members. Then the candidate gene *MVK* was further validated in the probands of an additional 57 DSAP pedigrees and 25 sporadic cases, 13 additional heterozygous mutations in *MVK* were identified. None of these mutations were detected in 676 unrelated ethnically matched controls and they were all predicted to be deleterious, which strongly support that *MVK* is the pathogenic gene responsible for DSAP. Further functional studies in cultured primary keratinocytes suggest that *MVK* has a role in regulating calcium-induced keratinocyte differentiation and could protect keratinocytes from apoptosis induced by type A ultraviolet radiation. The example demonstrated here employs the family-based strategy that is widely used for Mendelian disease gene discovery.

Table 7.5 Filtering steps of pathogenic mutations for the demo case

Filters	I:2	II:4	I:2+II:4	ANNOVAR	GERP[c]	ANNOVAR+GERP
NS/SS/I[a]	7,326	7,209	4,166	-	-	-
Not in dbSNP	972	958	447	-	-	-
Not in dbSNP or HapMap exomes	708	709	319	-	-	-
Not in dbSNP, HapMap exomes or 1000 Genomes	537	558	245	-	-	-
Not in dbSNP, HapMap exomes, 1000 Genomes or the control(II:1)	264	270	104	54	61	44
Located within linkage regions[b]	4	3	3	1(*MVK*)	2(*MVK, SELPLG*)	1(*MVK*)

[a]NS/SS/I refer to non-synonymous/splice-site/indel mutations.
[b]Candidate genomic regions identified by previous linkage analyses.
[c]GERP score greater than 2.

Challenges of exome sequencing in clinical applications

It is well known that clinical application of exome sequencing requires a perfect accuracy of variant identifications. Unfortunately, current NGS tools do not perform perfectly well in this realm. As an example, we tested four SNP calling methods, GATK (DePristo *et al.*, 2011), SOAPsnp (Li *et al.*, 2009b), SNVer (Wei *et al.*, 2011), GNUMAP algorithm (Clement *et al.*, 2010), to evaluate the 100X mean coverage exome data using default parameters. As shown in Fig. 7.7, although the majority of SNPs for one exome could be detected by all methods, there is poor concordance among different methods and a fraction of potentially authentic SNPs are missed and identified by individual method-specific pipelines.

For indel calling, three methods, including GATK, SAMTools (Li *et al.*, 2009a), SOAPindel (Li *et al.*, 2013), were used to conduct the comparisons under the default parameters. As expected, there was less consistency among different indel callers than that for SNP callers, which indicated the difficulty in identifying short indels through NGS tools (Fig. 7.8). These studies demonstrate that each pipeline alone will introduce some false positives and false negatives, highlighting the potential challenge for applying exome

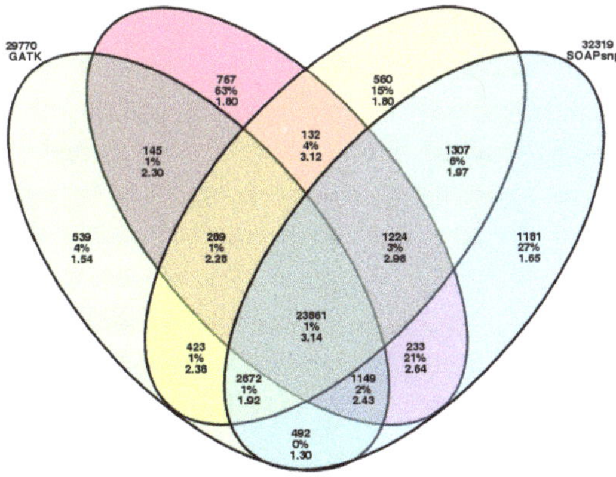

Figure 7.7 Venn diagram showing concordances and differences among different SNP calling pipelines. The figures in each cell refer to the number of SNPs, percentage of novel SNPs (compared with dbSNP135) and Ti/Tv ratio, respectively.

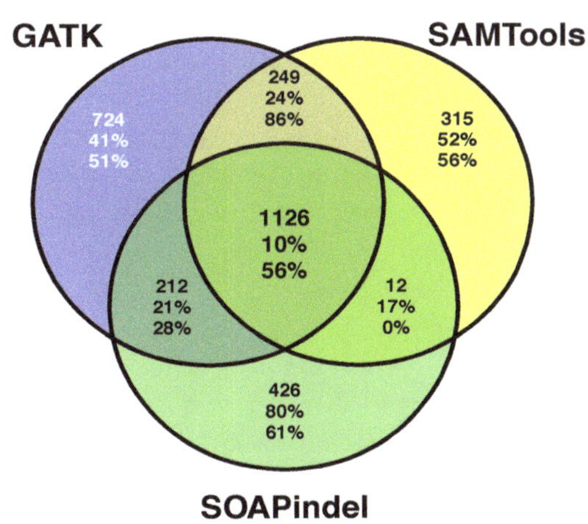

Figure 7.8 Venn diagram showing consistencies and differences among three popular indel calling pipelines. The numbers in each cell refer to indel number, the percentage of novel indels (compared with dbSNP135) and frameshift rate, respectively.

sequencing in clinical diagnosis and application with current NGS tools. Very recently, O'Rawe *et al.* extended the number of SNP calling pipelines for comparison to five and discovered that novel variants had much lower concordance among different pipelines than known variants (O'Rawe *et al.*, 2013), highlighting the challenge for accurately detecting rare variants by current NGS-based computational methods.

Optimization of variants calling may reduce some false positives (Reumers *et al.*, 2012), but at the same time also increase false negatives. The challenge of obtaining accurate variant calls from NGS data is substantial and it is mainly due to the inherent characteristics of NGS data itself. Because of these challenges, we believe a general analysis framework for NGS data that achieves consistent and accurate results from a wide range of experimental design options including diverse sequencing machinery and distinct sequencing approaches should be developed and standardized. As sequencing technologies, downstream mapping and variant calling algorithms improve, the informatics challenge is shifting from data production to data analysis. With the widespread adoption of ever-larger NGS generated datasets, the utility of NGS technology will only be fully realized when coupled with appropriate data mining tools to extract biologically meaningful information.

Applications in human disease research

As an efficient and cost-effective method, exome sequencing exhibits unique power to study human diseases. First, the discoveries of previous studies have proven that the strategy of focusing on protein-coding sequences is effective. Most causal variants underlying Mendelian disorders have altered protein-coding sequences. Second, many rare, protein-altering variants are predicted to be functionally important. Thus, exome is increasingly considered highly enriched for causal variants responsible for various human diseases.

However, this method exhibits its own limitations. First, the coverage of interested region is not complete for most kits. While enrichment kits have been much improved in recent years, the effective coverage for exome bases called at present is about 95%. Second, it is difficult to identify copy number variants (CNV) and assess other structure variants (SVs) using the current kits. This is because the false positive rate of CNV identified by exome sequencing is very high, and therefore the results are unreliable. At present the types of identified variants by exome sequencing are largely limited to SNPs and small indels. However, in consideration of the advantages and limitations, exome sequencing is still the best approach to identify most variants that cause alterations of protein sequences. As one of the most cost-effective NGS applications, exome sequencing has been rapidly adopted by scientists to study the pathogenesis of human diseases. It has been used to identify novel variants in normal individuals and patients associated with both rare and common disorders, such as Mendelian disorders, complex diseases and cancers.

Normal individuals and populations

The human exome was first analysed and characterized in normal individuals. In 2008, Ng *et al.* systematically assessed the genetic profile of the exome of a human individual, whose genome was previously sequenced to 7.5-fold coverage by traditional Sanger dideoxy technology (Levy *et al.*, 2007). The analysis found that the majority of the coding variants in this individual seemed to be functionally neutral, while approximately 10,400 non-synonymous single nucleotide polymorphisms (nsSNPs) were identified. About 1,500 nsSNPs were considered rare in the human population, and half of the approximately 700 coding indels caused insertions/deletions of amino acids rather than frameshifts in the corresponding proteins. This research suggests that many rare variants and non-SNP variants will be identified with more sequenced genomes in the future. Furthermore, a specific approach to analyse the coding variations in humans is presented by integrating multiple bioinformatics methods to pinpoint possible functional variation, which is a meaningful guidance for future researches on the variation in human exomes (Ng *et al.*, 2008).

Some interesting discoveries have been revealed by exome sequencing at the population level. For example, the Tibetans can live and work comfortably in high altitude environments (i.e. the Tibetan Plateau) while people living in low altitudes often experience high altitude sickness if they visit the Tibetan Plateau. The main reason why Tibetans are adapted to the high altitude was revealed in 2010 by exome sequencing (Yi *et al.*, 2010). In that study, researchers at BGI (formerly known as Beijing Genomics Institute) in China sequenced the exomes of 50 Tibetans with the NimbleGen 2.1M exon capture array to an average coverage of 18-fold through NGS. The exome result was then compared with exome sequences of 40 Chinese Han individuals living at low altitude (less than 50 m above sea level), as well as to 200 Danish exomes. The results revealed that the strongest selection signal came from the endothelial Per-Arnt-Sim (PAS) domain protein 1 (*EPAS1*), a transcription factor involved in response to hypoxia. One SNP at *EPAS1* associated with erythrocyte abundance exhibited a 78% frequency difference between Tibetan and Han samples, which was the fastest allele frequency change observed at any human gene to date. Due to the discovery of population-specific variants, the research also confirmed the effectiveness of exome sequencing researches at the population level.

After the successful application of exome sequencing at the population level, BGI subsequently accomplished another large-scale population research by this approach. An exciting discovery that coding regions harbour a larger proportion of low-frequency deleterious mutations was revealed by this research based on 200 human exomes. Li *et al.* used NimbleGen 2.1M exon capture array with an average 12-fold sequencing depth to investigate 200 Danes (Li *et al.*, 2010). 121,870 SNPs were identified in the sample population, including 53,081 coding SNPs (cSNPs). Through a rigorous statistical method for SNP calling and an estimation of allelic frequencies based on population data, the low-frequency allele spectrum of cSNPs was obtained with the range of minor allele frequency (MAF) between 0.02 and 0.05. This finding was different from previous viewpoints that ignored the low-frequency variants. The results also suggested that low-frequency variants could be very informative and play important roles in human fitness. Furthermore, a 1.8-fold excess of deleterious non-synonymous cSNPs over synonymous cSNPs was discovered in the low-frequency range. This excess was more evident for X-linked SNPs, suggesting that deleterious substitutions were primarily recessive. These studies indicate that exome sequencing is a powerful tool to reveal genetic variations underlying the diversity among individuals and to characterize the distribution of allele frequency in human populations.

Mendelian disorders

Mendelian disorders are genetic disorders, including autosomal recessive, autosomal dominant, X-linked dominant, X-linked recessive and Y-linked disorders. Mendelian disorders are typically caused by mutations in single genes and the inheritance pattern of Mendelian disorders in a family can be revealed by pedigree analysis. But the traditional approaches of linkage analysis and positional cloning require large size of patients and a family lineage with multiple generations. Such requirements limit the pace of disease gene discovery for researchers. Although many genes have been found to be associated with a variety of diseases, few have been confirmed as disease-causing (causative) genes. However, the pace of disease gene discovery is accelerating by the application of exome sequencing. As of 16 August 2013, the Online Mendelian Inheritance in Man (OMIM) database listed 7,481 known or suspected Mendelian diseases, with 3,866 (52%) of these having an identified molecular basis. The strength of exome sequencing lies in the wide range of applications. At present, exome sequencing can be applied to a relatively small number of affected individuals, in which the large amount of sequencing data are filtered to reduce the number of candidate genes to a single or a few with a high degree of confidence. The general approach is to assess the novelty of candidate variants by filtering the variants against available polymorphisms in public databases (e.g. dbSNP and 1000 Genomes Project) and/or those found in a set of unaffected individuals (controls), the detailed workflow has already been described in Fig. 7.6. Since NGS technologies have higher base calling error rates than Sanger sequencing, the candidate variants after filtration need to be validated by Sanger sequencing to minimize false calls.

Although the most common research strategy for studying Mendelian disorders using exome sequencing might be the family-based model in which affected individuals (cases) and unaffected individuals (controls) from the same family are selected for analyses, most of those successful studies have, to varying extents, relied on comparisons with exome sequences and variants that are found in a small number of unrelated affected individuals to find rare alleles or novel alleles in the same gene shared among cases (Bamshad *et al.*, 2011). Ng *et al.* first demonstrated the feasibility of using exome sequencing to identify disease-causing gene in Mendelian disorders with a small cohort of unrelated, affected individuals (Ng *et al.*, 2009). Twelve human exomes were sequenced after capturing by the Agilent 244K arrays, including eight normal individuals selected from the HapMap project and four unrelated individuals with the Freeman–Sheldon syndrome (FSS), a rare dominantly inherited disorder. The identified functionally important variants were filtered step by step against dbSNP database and eight HapMap control exomes, leaving *MYH3* as the only candidate gene in which each affected individual had at least one novel functional mutation. The *MYH3* gene had already been shown to cause FSS by conventional candidate gene approach. Using FSS as a proof-of-concept, this study aimed to develop a strategy for subsequent disease-causing gene identification based on exome sequencing. The results also suggest that this strategy can be extended to complex diseases with larger cohort of samples by appropriately weighting non-synonymous variants with predicted functional impact.

Enlightened by the successful design of using a small number of unrelated cases, disease-causing genes underlying many Mendelian disorders have been gradually identified by exome sequencing. For example, Ng *et al.* demonstrated the first successful application of exome sequencing to identify the gene in a rare Mendelian disorder of unknown cause, the Miller syndrome (MIM%263750) (Ng *et al.*, 2010). Four affected individuals in three independent families were studied by exome sequencing with an average coverage of 40X. Filtering against dbSNP database and eight HapMap exomes, *DHODH*, the gene encoding a key enzyme in the pyrimidine *de novo* biosynthesis pathway, was detected as the candidate. Then, mutations in *DHODH* were confirmed by Sanger sequencing to be present in three additional families with Miller syndrome. This research further confirms that exome sequencing of a small cohort of unrelated affected individuals is a powerful, efficient strategy for identifying the genes underlying rare Mendelian disorders.

However, the strategies using unrelated cases and/or family cases cannot be used to identify causative mutations underlying all Mendelian disorders. That is because some causative mutations are not inherited from parents, but only exist in children owing to a mutation in the germ cell (egg or sperm) of one of the parents or in the fertilized egg itself. This kind of mutation is called a *de novo* mutation and is present only in the affected new member of the family and not in the somatic cells of the parents. Exome sequencing of parent–child trios is extremely efficient to identify *de novo* coding mutations, in which parents and children in the same family are involved. This trio-based strategy may be particularly suitable for gene discovery in rare syndromes or disorders for which most cases are sporadic (that is, the parents are unaffected). The distinct difference between trio-based and family-based strategies is that unaffected parents are regarded as control in trio-based strategy while unaffected siblings or parents can both act as control in family-based strategy. As a well-justified strategy for revealing *de novo* mutations, the trio-based strategy of exome sequencing has accelerated the discoveries of genetic causes responsible for rare Mendelian disorders. For example, Hoischen *et al.* identified *de novo* mutations associated with the Schinzel–Giedion syndrome by trio-based exome sequencing of four affected individuals (cases) (Hoischen *et al.*, 2010). Heterozygous *de novo* mutations in *SETBP1* were detected in all four cases, and further confirmed in eight additional cases by Sanger sequencing. The same strategy was employed to discover the *de novo* nonsense mutations in *ASXL1* underlying Bohring–Opitz syndrome (Hoischen *et al.*, 2011). Moreover, Gibson *et al.* used trio-based exome sequencing to analyse two families affected by the Weaver syndrome, in which one of the original families was reported in 1974. Two different *de novo* mutations were identified in the enhancer of zeste homologue 2 (*EZH2*) by filtering rare mutations in the affected probands against the parental mutations. A third *de novo* mutation in *EZH2* was identified by Sanger sequencing in a third classically-affected proband. Taken together, *de novo* mutations in *EZH2* were proven to cause Weaver syndrome (Gibson *et al.*, 2012).

Due to the social significance and impact of Mendelian disorders, BGI launched the 1000 Mendelian Disorders Project in May 2010 to collaborate with researchers around the world to reveal causative genes underlying Mendelian disorders. The project first reported the identification of *TGM6* as a novel causative gene of spinocerebellar ataxia (Wang, J.L., *et al.*, 2010). Researchers sequenced the whole exome of four patients in a Chinese four-generation spinocerebellar ataxia family and identified a missense mutation, c.1550T>G transition (L517W), in exon 10 of *TGM6*. After validation by linkage analysis, *TGM6* was confirmed as the causative gene by identifying another missense mutation c.980A>G transition (D327G) in exon 7 of *TGM6* in an additional spinocerebellar ataxia family. The conclusion was further confirmed by the absence of both mutations in 500 normal unaffected individuals of matched geographical ancestry. Shortly after the first publication, another successful application of exome sequencing was reported to identify a gene potentially responsible for high myopia, an autosomal dominant Mendelian disorder (Shi *et al.*, 2011). Two affected individuals from a Han Chinese family with high myopia were captured by NimbleGen 2.1M array and sequenced by NGS with an average coverage of 30X. The shared genetic variants of two patients in the family were filtered against the 1000 Genomes Project and the dbSNP131 database. A mutation A672G in the zinc finger protein 644 isoform 1 (*ZNF644*) was identified as being related to the phenotype of this family. After sequencing the exons in the *ZNF644* gene in 300 sporadic cases of high myopia, five additional mutations (I587V, R680G, C699Y, 3′UTR+12 C>G and 3′UTR+592 G>A) were identified in 11 different patients, all of which were absent in 600 normal ethnical matched controls. Since *ZNF644* was predicted to be a transcription factor that might regulate genes involved in eye development, *ZNF644* was inferred to be a causal gene for high myopia in a monogenic form. This study showed that exome sequencing of affected individuals in one family is an cost-effective method for identifying causative genes underlying rare Mendelian disorders. Since its launch, the 1000 Mendelian Disorders Project has reported dozens of successful cases of exome sequencing in disease gene discovery (Wang, J.L., *et al.*, 2010, 2011; Liu *et al.*, 2011; Shi *et al.*, 2011; Chiang *et al.*, 2012; Guo *et al.*, 2012a; Lin *et al.*, 2012; Zhang *et al.*, 2012a,b, 2013; Quadri *et al.*, 2013).

Complex diseases

Different from Mendelian disorders caused by a single-mutated gene, complex diseases are mostly caused by combinations of multiple genetic and environmental factors. Thus, the interplay between genetic and environmental factors in complex diseases remains a great challenge to researchers. It is widely recognized that complex diseases do not obey the standard Mendelian patterns of inheritance. At present, genome-wide association studies (GWAS) have been demonstrated to be a powerful approach to identify genetic basis underlying complex diseases. Until now, GWAS have identified hundreds of thousands of common genetic variants associated with complex diseases and have provided valuable insights into the complexities of their genetic architecture. However, in almost all cases, these loci collectively account for only a small fraction of the observed heritability of complex diseases. Moreover, it is unclear to what extent rare alleles (minor allele frequency (MAF) $\leq 5\%$) might contribute to the heritability of complex traits, raising the question of finding the 'missing heritability' of complex diseases (Manolio *et al.*, 2009). An individual's genetic disease risk arises from the collection of common variants inherited from distant ancestors (small effect), as well as rare variants from more recent ancestors (potentially larger effect) and *de novo* mutations (Lupski *et al.*, 2011). However, strong evidence suggests that rare variants of larger effect are responsible for a substantial portion of complex human diseases (McClellan and King, 2010). The collection of older common variants may have less influence on an individual's disease susceptibility than that of recently arisen rare variants and *de novo* mutations. Therefore, whole genome re-sequencing rather than genotyping is needed to identify these recently arisen rare alleles and *de novo* mutations. Compared with Mendelian disorders, the contribution of exome sequencing to our understanding of complex diseases has been limited, although whole genome sequencing and target re-sequencing of GWAS loci have identified several low-frequency and rare variants for complex diseases and traits (Holm *et al.*, 2011; Rivas *et al.*, 2011; Sulem *et al.*, 2011; Jonsson *et al.*, 2012). However, we believe exome sequencing is among the most cost-effective NGS-based approaches to reveal causative coding variants for complex diseases.

Similar to that for Mendelian disorders, the trio-based and family-based strategies can be used to respectively identify *de novo* and rare variants in multiple affected individuals with a common trait in families. It is hypothesized that *de novo* mutations may play a great role in the genetics of common neurodevelopmental and psychiatric diseases. Vissers *et al.* first reported a trio-based study of exome sequencing of affected offspring and unaffected parents to test this *de novo* mutations hypothesis. By sequencing the exomes of ten case-parent trios, *de novo* pathogenic mutations were identified in six of these cases with unexplained mental retardation. This study demonstrated a *de novo* paradigm for mental retardation and highlighted the power of trio-based exome sequencing in understanding the genetic basis for common mental diseases (Vissers *et al.*, 2010). Inspired by this pioneering study, O'Roak *et al.* used trio-based exome sequencing to identify new candidate genes for autism spectrum disorders (ASDs) (O'Roak *et al.*, 2011). They sequenced the exomes of 20 individuals with sporadic ASD (case) and their parents (control). 21 *de novo* mutations were identified, 11 of which were protein-altering coding variants. Among them, potentially causative *de novo* mutations were identified in 4 out of 20 probands (in *FOXP1*, *GRIN2B*, *SCN1A* and *LAMC3*), particularly among more severely affected individuals. This research suggested that *de novo* mutations might contribute substantially to the genetic aetiology of ASDs. The family-based strategy of exome sequencing is also used to study complex diseases that are caused by inherited mutations in families. Musunuru *et al.* performed exome sequencing in two family members affected by combined hyperlipidaemia, marked by extremely low plasma levels of low-density lipoprotein (LDL) cholesterol, high-density lipoprotein (HDL) cholesterol, and triglycerides (Musunuru *et al.*, 2010). After variants calling and filtering, they found that these two cases were compound heterozygotes for two distinct nonsense mutations in *ANGPTL3* (encoding the angiopoietin-like 3 protein). They further showed that the nonsense mutations in *ANGPTL3* affected LDL cholesterol and triglyceride phenotypes in a manner dependent on the gene dosage while were recessive with respect to HDL cholesterol. Their finding of *ANGPTL3* mutations highlights a role for this gene in LDL cholesterol metabolism in humans. Furthermore, Lyon *et al.* applied exome sequencing to study several affected individuals with attention deficit/hyperactivity disorder (ADHD) in a pedigree (Lyon *et al.*, 2011). Some rare candidate variants that predispose to ADHD were identified in the affected family members. These studies demonstrated that exome sequencing could reveal inherited variants that might play a role in familial complex diseases.

However, the most widely used strategy for complex disease research is the case–control design. In this design, the allele frequencies in a group of unrelated affected individuals (cases) with the disease of interest are compared to those in a disease-free comparison group composed of unrelated normal individuals (controls) (Pearson and Manolio, 2008). This strategy requires large cohorts of cases and controls to identify disease-related mutations, especially for rare and low-frequency mutations. According to a simulation on the association power, at least 2,000 samples are needed to find low-frequency variants ($0.02 \leq MAF < 0.05$) predisposing to complex diseases (Table 7.6). The probability of identifying causative mutations increases with sample size as expected. Compared with common

Table 7.6 The sample size requirements for detecting statistically significant variant–disease associations in case–control studies based on simulation

Case/control	Option 1	Option 2	Option 3	Option 4
Sample size	100/100	300/300	500/500	1000/1000
Statistical power (Prevalence >=1%, RR >=2, depth=16)	≥8.98% (MAF≥0.02)	≥54.38% (MAF≥0.02)	≥77.20% (MAF≥0.02)	≥96.42% (MAF≥0.02)
	≥43.53% (MAF≥0.05)	≥90.76% (MAF≥0.05)	≥98.32% (MAF≥0.05)	≥99.96% (MAF≥0.05)

The full names and interpretation are represented as follows: prevalence means the proportion of a population found to have a disease at a given time; RR (risk ratio) represents the contribution of the mutation to disease occurrence; and MAF (minor allele frequency) is the frequency of the SNP's less frequent allele in a given population.

and low-frequency mutations, rare mutations (MAF < 0.01) require even larger sample sizes for detecting robust association signals. Under a fixed sequencing budget, Li *et al.* demonstrated that sequencing many individuals at low depth outperforms the strategy of sequencing fewer samples at high depth in terms of sensitivity of variant discovery and statistical power for association mapping, indicating that low-coverage sequencing is an attractive strategy for complex disease association studies (Li *et al.*, 2011). Pasaniuc *et al.* also demonstrated the benefits of low-coverage sequencing-based GWAS by showing that extremely low-coverage sequencing coupled with genotype imputation from 1000 Genomes Project reference panel could capture almost all common and low-frequency variants, and substantially increase statistical power for GWAS compared with traditional array genotyping (Pasaniuc *et al.*, 2012). To further investigate how many samples would be enough to achieve sufficient power to associate rare variants with disease phenotypes in a sequencing-based GWAS, we made simulation studies to evaluate the association power under different combinations of sample size and MAF level (Fig. 7.9). As can be seen from Fig. 7.9, at least 5,000 samples will be needed to have > 80% power to detect a causal rare variant (MAF of 0.5%) with a threefold effect (OR = 3) at the genome-wide significance level of 1×10^6. Our simulation results thus have important implications for designing a powerful sequencing-based GWAS to detect rare disease-associated variants.

The first generation large-scale exome sequencing for complex disease research was launched by nine Danish university centres/institutes and BGI from China in early 2008. This was part of the broad Danish-Sino initiative aimed at understanding the human genome and the gut microbiome in order to improve metabolic and cardiovascular health of the at-risk population. This pioneering project was funded by the Lundbeck Foundation Centre for Applied Medical Genomics in Personalised Disease Prediction, Prevention and Care (LuCamp). Its purpose was to identify common metabolic traits associated genes by exome sequencing of 1,000 patients with the combined at-risk metabolic phenotypes of visceral obesity, type 2 diabetes (T2D) and hypertension and 1,000 age- and gender-matched, glucose-tolerant, lean and normotensive individuals (http://www.lucamp.org). In this study, medium-depth (8X) whole exome sequencing was performed on 2,000 Danish individuals in stage 1. Through rigorous data quality control, SNP detection and allele frequency estimation, and association test directly on the mapped reads, 70,182 SNPs with MAF >1% were identified, of which 995 SNPs showed nominal association with disease status. In stage 2, 16,192 SNPs, including all functional SNPs identified in exome sequencing, were selected for Illumina iSelect genotyping in 15,989 Danes to search for association with 12 metabolic phenotypes, 51 potential association signals were identified for one or more of eight metabolic traits. In stage 3, 45 coding SNPs showing potential association were further genotyped in up to 63,896 European individuals for replication. Meta-analyses of stage 2 and stage 3 results demonstrated robust associations for three common and low-frequency coding variants in *CD300LG* (fasting HDL-cholesterol: R82C, MAF 3.5%,

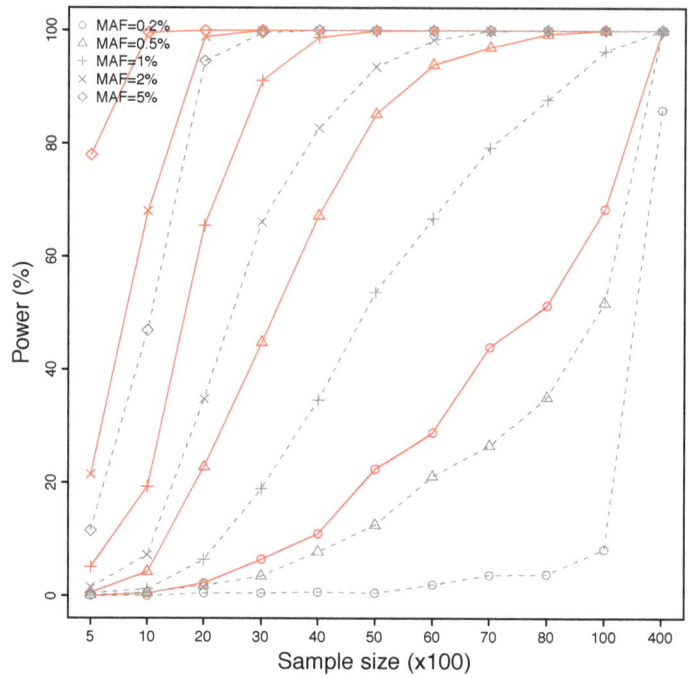

Figure 7.9 Power comparison under different sample sizes and MAF ranges, given sequencing depth=8X, disease prevalence Kp=0.001, e=0.002 (sequencing error) and alpha=1×10^{-6} (statistical significance). Each curve shows the power (y-axis) to detect a causative SNP with a specified MAF and a specified OR (effect size) (red: OR=3, gray: OR=2).

$P=8.5\times10^{-14}$), *COBLL1* (type 2 diabetes: N939D, MAF 12.5%, OR = 0.88, $P=1.2\times10^{-11}$) and *MACF1* (type 2 diabetes: M2290V, MAF 23.4%, OR = 1.10, $P=8.2\times10^{-10}$). This well-powered three-stage association study systematically assessed the contribution of common and low-frequency functional coding variants to the genetic susceptibility of common metabolic phenotypes such as T2D, obesity and hypertension, and concluded that coding polymorphisms with MAF above 1% do not seem to have particularly high effect sizes on the measured metabolic traits, which could serves as an indication of the utility of exome sequencing in complex metabolic traits (Albrechtsen *et al.*, 2013). Although this study did not investigate rare coding variants predisposing to metabolic traits, it still represented a significant effort towards understanding the genetic determinations of complex traits. The rare coding variants were not investigated because (i) large-scale high coverage exome sequencing was not affordable when the project was initiated and thus genotypes for each individual can not be accurately called, and (ii) the lack of robust Danish reference panel available for imputing the genotypes of rare variants from this low-coverage exome sequencing data. In the near future, even large-scale deep exome sequencing or low-pass whole genome sequencing coupled with imputation based on local well-powered reference panel will provide sufficient power to enable burden test analyses of the combined impact on complex phenotypes of multiple rare and low-frequency variants in a given gene locus or in other functional units such as a biologically relevant pathway.

Recently, rare coding variants in known risk genes were demonstrated to have negligible impact on the missing heritability for autoimmune disease phenotypes, which reduces the passion for conducting large-scale whole exome sequencing projects in common autoimmune diseases (Hunt *et al.*, 2013). However, the research community is still awaiting the completion of well-powered unbiased large-scale exome sequencing studies to demonstrate the contribution of rare coding mutations to complex phenotypes. With reduced costs of exome sequencing and increasing numbers of publicly available control exomes, the case–control strategy will be increasingly used by the research community to study causative genes of complex diseases, especially those that could be used for large populations in clinics.

In conclusion, several strategies of exome sequencing can be used to identify disease-causing variants, in which the workflow of exome sequencing is identical or very similar. However, the choice of strategies for finding causal alleles depends on many factors, including the inheritance mode of a trait, the pedigree or population structure, the pathogenesis of phenotype owing to de novo or inherited variants, etc. Together, these factors influence the sample size that is required to provide adequate statistic power to detect disease-causing variants.

Cancers

The application of NGS has greatly advanced the progress of cancer genome research owing to its high-throughput advantage. The major challenge of cancer genome research is to identify driver mutations that confer growth advantage to cancer cells (Stratton *et al.*, 2009). As the mutations causing cancer are heterogeneous, which may include SNP, indel, copy number variation (CNV) and structural variation (SV), there are challenges of using NGS to detect all the cancer driver mutations. For example, although CNV can be detected by exome sequencing to some extent, the high rate of false positives makes the results unreliable. Thus, exome sequencing is seldom used to detect CNV. At present, SNPs and small indels in coding regions are almost the exclusive type of driver mutations identified by exome sequencing.

Historically, exome sequencing was first applied as a traditional PCR-based sequencing method in cancer genomics research (Sjoblom *et al.*, 2006). Its wide application based on NGS technologies has promoted its use for discovering disease genes associated with many cancer types, such as ovarian, bladder, lymphoma and leukaemia cancers (Jones *et al.*, 2010; Gui *et al.*, 2011; Pasqualucci *et al.*, 2011; Yan *et al.*, 2011). The results of these studies on the distinct subtypes of cancer are beginning to reveal the potential mechanisms underlying specific cancers. The first report of NGS-based exome sequencing in cancer genomics analysed exome sequences of 8 ovarian clear cell carcinomas (OCCCs) and compared them with those of normal cells from the same patients (Jones *et al.*, 2010). After enrichment by the Agilent SureSelect system, DNA was sequenced using the Illumina GAIIx platform with an average coverage of 84X. 268 somatic mutations in 253 genes were detected in the 8 tumours, of which 237 (88%) were confirmed by Sanger sequencing. Mutations in four genes were identified in at least two tumours from the discovery stage. In the validation stage, the four frequently mutated genes were further validated using the tumour and normal tissues of an additional 34 OCCCs for screening the prevalence of these mutations by PCR amplification and Sanger sequencing. In total, the *PIK3CA*, *KRAS*, *PPP2R1A*, and *ARID1A* mutations were identified in 40%, 4.7%, 7.1%, and 57% of the 42 tumours, respectively. This study demonstrated that *PPP2R1A* was a novel oncogene, and *ARID1A* was a novel tumour suppressor gene (Fig. 7.10). The case–control design and the analysis pipeline in this research serve as the mainstream strategy for the subsequent exome sequencing based cancer genomics researches.

As shown above, the case–control design in cancer genomics is different from those in Mendelian and complex diseases genomics. For Mendelian and complex diseases, peripheral blood (PB) of affected (patients) and unaffected individuals (normal people) is the typical material for analyses from the case and control, respectively. However, in cancer genomics, PB is not the only material used. That is because cancer can be caused by both germline and somatic mutations. Thus, the choice of case and control depends on the type of mutations. On the one hand, germline mutations are inherited from parents. To detect germline mutations, the case–control design in cancer genomics is identical to that in Mendelian and complex diseases. This strategy is particularly used to study familial inherited cancers. For example, hereditary pheochromocytoma (PCC) is often caused by germline mutations. By exome sequencing of three unrelated individuals with hereditary PCC (cases), mutations were identified in *MAX*, the MYC associated factor X gene. In a follow-up study of 59 selected PCC cases,

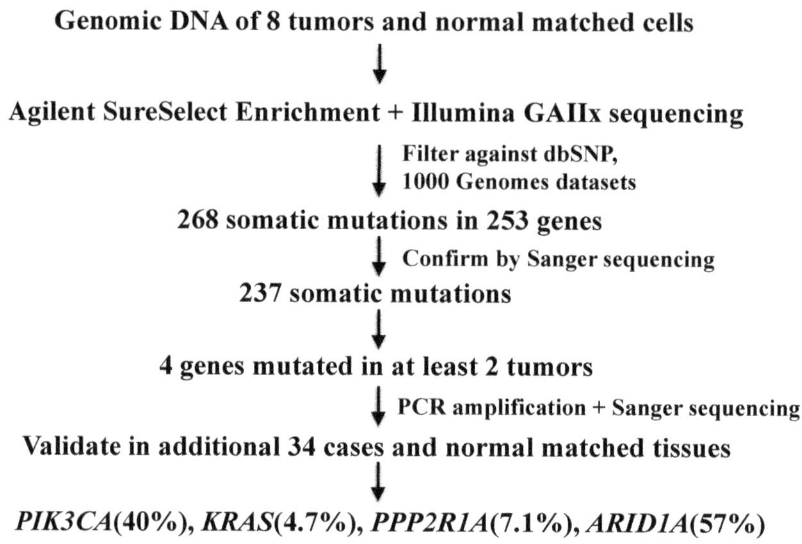

Figure 7.10 The analysis pipeline for identifying somatic mutations in ovarian clear cell carcinoma through exome sequencing.

five additional *MAX* mutations were identified, which indicated an association with malignant outcome and preferential paternal transmission of *MAX* mutations. Further functional and network analysis suggested that loss of *MAX* function was correlated with metastatic potential (Comino-Mendez *et al.*, 2011). On the other hand, somatic mutations arise after birth as non-inherited mutation. To detect somatic mutations, the choice of case and control depends on the tumour type. For solid tumour, e.g. liver, lung and bladder cancer, tumour tissue is selected as the case while normal matched tissue or PB of the same patient could either serve as the control to filter out germline variants in cases. For example, Guo *et al.* performed exome sequencing of ten clear cell renal cell carcinomas (ccRCCs) with the average depth of 127X in the discovery screen, followed by targeted re-sequencing of ~1,100 candidate genes in 88 additional ccRCCs in the prevalence screen, 23 genes somatically mutated at elevated frequencies were identified in the total 98 matched pairs of tumour tissues (cases) and normal tissues (controls), 12 of which were first reported to be associated with this tumour. Further pathway analysis revealed that frequent mutations were involved in the ubiquitin-mediated proteolysis pathway (UMPP), suggesting the potential contribution of UMPP to ccRCC tumorigenesis (Guo *et al.*, 2012b). For haematological tumour, e.g. leukaemia, somatic mutations are contained in the blood. PB or bone marrow is accordingly selected as the case, instead of control. And normal skin tissue of the same patient usually serves as the control, but PB during complete remission after standard chemotherapy could also be used as the control if other normal tissues are not available. For example, nine pairs of the M5 subtype of acute myeloid leukaemia (AML-M5) were studied by exome sequencing with the average depth of 100X. Bone marrow at diagnosis was selected as case, and PB after complete recovery was treated as control. Confirmed by Sanger sequencing, 66 somatic mutations were identified in 63 genes. Through validation of 63 genes in a large cohort of AML-M5 patients, *DNMT3A*, encoding a DNA methyltransferase, was found mutated in 23 of 112 (20.5%) cases. Further functional analysis suggested a contribution of aberrant DNA methyltransferase activity to the pathogenesis of acute monocytic leukaemia (Yan *et al.*, 2011). It's now well established that somatic mutations are the main cause of tumorigenesis. Thus, the case–control design for identifying somatic mutations is the most frequently used strategy in cancer genomics research. In general, the case–control design in cancer genomics is complex due to the fact that high-quality source DNA is needed (Hudson *et al.*, 2010). To perform exome sequencing of cancer, the first step is to investigate whether the studied cancer is familial or not. Then, suitable samples are collected to serve as case and control and different types of tissue samples are obtained based on the tumour type. The quantity and quality of collected samples can also influence the research result.

Due to recent developments of experimental technologies, single-cell exome sequencing is showing great promise and exhibits a unique advantage to reveal the evolutionary process of cancer at the single-cell level. Different from typical exome sequencing for tissues described above, the objective here is to obtain and analyse a single cell from tumour tissue. The DNA from the isolated single cell is amplified by whole genome amplification (WGA) technologies to generate sufficient DNA for NGS. The subsequent steps of exome capture and sequencing are identical to tissue exome sequencing. Now, BGI has applied this approach to detect SNV and Indel (Fig. 7.11). Due to its high resolution, single-cell exome sequencing can be widely applied to reveal the detailed mechanisms of oncogenesis and development. For example, this approach can be used to detect the genetic diversity among cells within the same cancer tissues. Based on the genetic differences among the subclones of cells in the tissues, the process of cancer metastasis could be precisely reconstructed. Thus, single-cell exome sequencing is expected to reveal the subclones carrying metastatic mutations and identify clinical biomarkers for metastasis treatment.

Recently, two papers using single-cell exome sequencing were published by BGI in Cell journal to exhibit its power in cancer research. Hou *et al.* sequenced 90 cells from a *JAK2*-negative myeloproliferative neoplasm patient, 58 cells passed quality control criteria and were used for data analyses. Subsequent population genetic analysis based on the somatic mutant allele frequency spectrum indicated the monoclonal evolution in this neoplasm (Hou *et al.*, 2012). Furthermore, essential thrombocythemia (ET)-related candidate mutations such as *SESN2* and *NTRK1* were identified, indicating their potential involvement in neoplasm progression. The research established the power of single-cell sequencing and opened the door for detailed analyses of a variety of tumour types. In the accompanying paper, the research team of BGI subsequently applied the established single-cell sequencing method to study clear cell renal cell carcinoma (ccRCC) (Xu *et al.*, 2012). A ccRCC tumour and its adjacent kidney tissue were analysed by single-cell exome sequencing, which delineated a detailed intratumoral genetic landscape at the single-cell level. The result showed that the tumour did not contain any significant clonal subpopulations. In addition, different mutation spectrums were discovered within the population by identifying mutations that had different allele frequencies. The research demonstrated the genetic complexity of ccRCC beyond previous expectation, and provided new ways to develop more effective cellular targeted therapies for individual tumours.

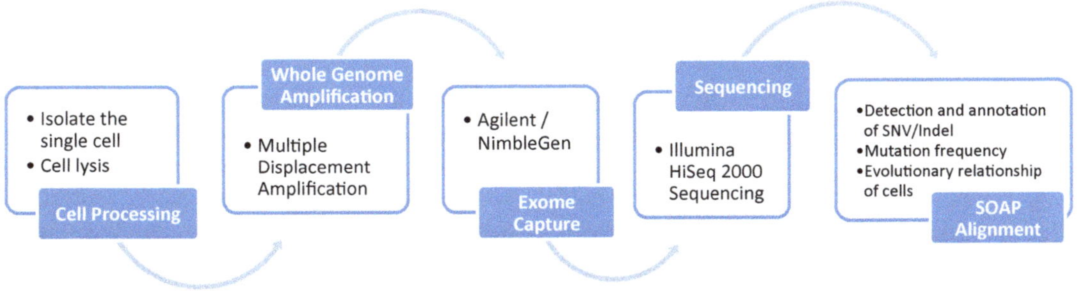

Figure 7.11 The workflow of single-cell exome sequencing.

Perspective

Exome sequencing is expected to be a routine technique for academic and clinical use, such as the Sanger sequencing technique. Thus, standardized protocols and analysis pipelines should be developed to improve the accuracy of variation detection and minimize false positive and false negative results. In the area of Mendelian disease research, the publications will always report the positive results, while it is unclear how many studies fail to identify the casual gene by exome sequencing. Potential reasons for failing to detect a compelling candidate gene include (i) the disease under investigation is genetically heterogeneous and that not all affected individuals carry the same mutation(s); (ii) the complete catalogue of mutations can not be identified in the targeted exome due to the limitation of exome capture technology itself (missing rate < 5%); (iii) some or all causative mutations are not in the targeted exome, but in noncoding regulatory regions or unannotated genes; (iv) the variants are covered but not accurately detected due to the complexity of target region, e.g. CNV or complex genomic rearrangement etc.; (v) issue of false positive and false negative in variant detection due to sequence alignment and analysis software; and (vi) the number of datasets as controls is not large enough to filter out normal variants that are not related to the specific disease. Some causative variants may be filtered as normal due to the lack of proper controls and/or that the authentic mutation is contained in the controls, e.g. some clinical-relevant mutations deposited in dbSNP database. However, with the continuing improvement in sequencing and capture technologies, new statistical and computational methods for variant calling, as well as an increasing number of sequenced exomes available for comparison, all these technical problems can be overcome, which will increase the already impressive success rate of exome sequencing studies for Mendelian disorders.

For complex disease research, the advantage of whole exome sequencing is the ability to perform deep sequencing of the coding portions of the genome at a relatively affordable cost. Such an approach allows the detection of rare coding variants that may have profound impact on protein function. To systematically interrogate the impact of rare coding variants on complex traits and conduct the sequencing-based GWAS successfully, there are still several challenges that need to be addressed. First, current sequencing costs prohibit the adequately sized samples (> 10,000 samples) to be deeply sequenced in order to reach enough power. With sequencing cost continues to decline, large-scale well-powered deep exome sequencing will be carried out on the well-phenotyped cohorts by international collaborative effort. Second, statistical and computational methods for sequencing-based GWAS involving rare variants are not well documented and the development is still in its infancy (Bansal *et al.*, 2010; Nielsen *et al.*, 2011). Although some rare variant association burden test methods, such as CMC (Li and Leal, 2008), WSS (Madsen and Browning, 2009), KBAC (Liu and Leal, 2010), VT test (Price *et al.*, 2010) and C-alpha (Neale *et al.*, 2011), are proposed and shown to have good power in detecting association signals in some circumstance by simulation studies, their pros and cons for large-scale empirical exome sequencing data should be fully evaluated and well-established. Besides, these tests should be implemented in one software suite to be applied to the same study (Stitziel *et al.*, 2011). Further development in this direction should be a priority for complex disease genetic research. Third, sequencing data and subsequent analysis require additional computational resources, including both the storage and high-performance computing infrastructures, which is far beyond what is available to analyse conventional chip-based GWAS data. Thus, well-organized data file format needs be developed to store this kind of large-scale data, and parallel and distributed computing will also be needed to accelerate the pace of analysing the data.

Recent studies have shown that human populations harbour an abundance of rare functional variants (Nelson *et al.*, 2012). Large-scale population deep exome sequencing studies have also revealed that the vast majority of coding variants is rare and novel, with the rarer variants being more likely to be population-specific and their numbers increasing linearly with sample size (Tennessen *et al.*, 2012). In addition, most of these rare coding variants seemed to have emerged more recently than common variants, and that human disease genes contain a significantly higher proportion of recently arisen deleterious mutations than other genes (Fu *et al.*, 2013). In statistical association tests, rare variants have shown systematically different stratification in spatially structure populations compared with common variants and such stratifications can confound real genotype–disease association signals (Mathieson and McVean, 2012). One aim of the 1000 Genomes Project is to create maps of genetic variation across multiple populations for understanding the genetic contribution to human disease. However, rare variants are difficult to identify in small local populations due to the typically small number of individuals sequenced per population in most studies (Abecasis *et al.*, 2012). Thus, systematic large-scale sequencing across a range of different populations is necessary for detecting those disease-associated rare variants. To reach this goal, reference panel construction based on large-scale whole genome sequencing for specific populations is needed for implementing an effective sequencing-based GWAS. Indeed, several such efforts have been initiated. For example, BGI has collaborated with research scientists in Denmark to launch the Danish Pan-Genomes Project, which aims to build the baseline of Denmark population to facilitate disease research and drug development. In addition, BGI also collaborated with research groups from the Netherlands to initiate The Genome of the Netherlands (GoNL) project, as part of the Dutch National Biobanks. GoNL employs a whole genome sequencing approach to analyse a representative sample consisting of 250 trio-families from all provinces in the Netherlands. The project aims to characterize DNA sequence variation in the Dutch population and establish a reference panel for imputation in the available genome-wide association studies in Dutch and other cohorts to refine association signals and uncover population-specific variants (Boomsma *et al.*, 2013). Other large-scale whole genome sequencing projects, such as the UK10K (http://www.uk10k.org) and SardiNIA Projects (http://sardinia.irp.nia.nih.gov), are ongoing. Recently, BGI has also initialized The China National Genebank (CNGB) project funded by the government of China. This large-scale project will store not only the biological resources and the bioinformatic data, but also construct a powerful reference panel for the Chinese population for future medical researches. CNGB is committed to developing a biobank consortium across China and an expanding network worldwide to provide better support for information sharing and exchange of Biobank construction, omics data acquiring, and multi-omics scientific researches (http://www.nationalgenebank.org/en/index.html). It is estimated that at least one million individuals will be needed to construct a powerful reference panel for the Chinese Han population. Taken together, these reference panels for local populations could serve as the foundation for larger scale sequencing-based GWAS and ultimately understanding the rare genetic variants in human health and disease.

It has been shown that common disease-associated variants are often localized in regulatory regions of the genome (Maurano *et al.*, 2012). Furthermore, the ENCODE projects have revealed that ~80% of the human genome, particularly the parts containing not well-studied non-coding regions, are biochemically functional (Bernstein *et al.*, 2012). These studies raise the possibility that many rare non-coding variants may channel their effects on disease susceptibility through regulatory elements distributed across the genome. Thus, whole genome sequencing, instead of exome sequencing, will be needed to uncover the genetic basis for such complex diseases. With all the susceptible genetic factors identified, the success of complex trait genetic research will then be determined by our ability to interpret the data and assemble sufficiently large well-phenotyped clinical populations.

As for cancer research, certain driver mutations such as structural variants and large genomic rearrangements cannot be detected by exome sequencing. Detecting such mutations will require complete whole genome re-sequencing (Campbell *et al.*, 2010; Ding *et al.*, 2010; Pleasance *et al.*, 2010). In summary, as the price of sequencing continues to decrease, researchers will be able to afford sequencing a higher coverage of the exomes for a larger number of samples. With its routine use in mutation discovery, we believe exome sequencing, and eventually whole genome sequencing will be a particularly powerful tool for common clinical diagnosis and personalized medicine, directly contributing to the management and improvement of human health.

References

Abecasis, G.R., Auton, A., Brooks, L.D., DePristo, M.A., Durbin, R.M., Handsaker, R.E., Kang, H.M., Marth, G.T., and McVean, G.A. (2012). An integrated map of genetic variation from 1,092 human genomes. Nature *491*, 56–65.

Adzhubei, I.A., Schmidt, S., Peshkin, L., Ramensky, V.E., Gerasimova, A., Bork, P., Kondrashov, A.S., and Sunyaev, S.R. (2010). A method and server for predicting damaging missense mutations. Nat. Methods *7*, 248–249.

Albrechtsen, A., Grarup, N., Li, Y., Sparso, T., Tian, G., Cao, H., Jiang, T., Kim, S.Y., Korneliussen, T., Li, Q., *et al.* (2013). Exome sequencing-driven discovery of coding polymorphisms associated with common metabolic phenotypes. Diabetologia *56*, 298–310.

Asan, Xu, Y., Jiang, H., Tyler-Smith, C., Xue, Y., Jiang, T., Wang, J., Wu, M., Liu, X., Tian, G., *et al.* (2011). Comprehensive comparison of three commercial human whole-exome capture platforms. Genome Biol. *12*, R95.

Bamshad, M.J., Ng, S.B., Bigham, A.W., Tabor, H.K., Emond, M.J., Nickerson, D.A., and Shendure, J. (2011). Exome sequencing as a tool for Mendelian disease gene discovery. Nat. Rev. Genet. *12*, 745–755.

Bansal, V., Libiger, O., Torkamani, A., and Schork, N.J. (2010). Statistical analysis strategies for association studies involving rare variants. Nat. Rev. Genet. *11*, 773–785.

Bernstein, B.E., Birney, E., Dunham, I., Green, E.D., Gunter, C., and Snyder, M. (2012). An integrated encyclopedia of DNA elements in the human genome. Nature *489*, 57–74.

Blanchette, M., Kent, W.J., Riemer, C., Elnitski, L., Smit, A.F., Roskin, K.M., Baertsch, R., Rosenbloom, K., Clawson, H., Green, E.D., *et al.* (2004). Aligning multiple genomic sequences with the threaded blockset aligner. Genome Res. *14*, 708–715.

Boomsma, D.I., Wijmenga, C., Slagboom, E.P., Swertz, M.A., Karssen, L.C., Abdellaoui, A., Ye, K., Guryev, V., Vermaat, M., van Dijk, F., *et al.* (2013). The Genome of the Netherlands: design, and project goals. Eur. J. Hum. Genet. [Epub ahead of print.]

Campbell, P.J., Yachida, S., Mudie, L.J., Stephens, P.J., Pleasance, E.D., Stebbings, L.A., Morsberger, L.A., Latimer, C., McLaren, S., Lin, M.L., *et al.* (2010). The patterns and dynamics of genomic instability in metastatic pancreatic cancer. Nature *467*, 1109–1113.

Challis, D., Yu, J., Evani, U.S., Jackson, A.R., Paithankar, S., Coarfa, C., Milosavljevic, A., Gibbs, R.A., and Yu, F. (2012). An integrative variant analysis suite for whole exome next-generation sequencing data. BMC Bioinformatics *13*, 8.

Chiang, P.W., Wang, J., Chen, Y., Fu, Q., Zhong, J., Yi, X., Wu, R., Gan, H., Shi, Y., Barnett, C., *et al.* (2012). Exome sequencing identifies NMNAT1 mutations as a cause of Leber congenital amaurosis. Nat. Genet. *44*, 972–974.

Clement, N.L., Snell, Q., Clement, M.J., Hollenhorst, P.C., Purwar, J., Graves, B.J., Cairns, B.R., and Johnson, W.E. (2010). The GNUMAP algorithm: unbiased probabilistic mapping of oligonucleotides from next-generation sequencing. Bioinformatics *26*, 38–45.

Comino-Mendez, I., Gracia-Aznarez, F.J., Schiavi, F., Landa, I., Leandro-Garcia, L.J., Leton, R., Honrado, E., Ramos-Medina, R., Caronia, D., Pita, G., *et al.* (2011). Exome sequencing identifies MAX mutations as a cause of hereditary pheochromocytoma. Nat. Genet. *43*, 663–667.

Cooper, G.M., Goode, D.L., Ng, S.B., Sidow, A., Bamshad, M.J., Shendure, J., and Nickerson, D.A. (2010). Single-nucleotide evolutionary constraint scores highlight disease-causing mutations. Nat. Methods *7*, 250–251.

Danecek, P., Auton, A., Abecasis, G., Albers, C.A., Banks, E., DePristo, M.A., Handsaker, R.E., Lunter, G., Marth, G.T., Sherry, S.T., *et al.* (2011). The variant call format and VCFtools. Bioinformatics *27*, 2156–2158.

DePristo, M.A., Banks, E., Poplin, R., Garimella, K.V., Maguire, J.R., Hartl, C., Philippakis, A.A., del Angel, G., Rivas, M.A., Hanna, M., *et al.* (2011). A framework for variation discovery and genotyping using next-generation DNA sequencing data. Nat. Genet. *43*, 491–498.

Ding, L., Ellis, M.J., Li, S., Larson, D.E., Chen, K., Wallis, J.W., Harris, C.C., McLellan, M.D., Fulton, R.S., Fulton, L.L., *et al.* (2010). Genome remodelling in a basal-like breast cancer metastasis and xenograft. Nature *464*, 999–1005.

Durbin, R.M., Abecasis, G.R., Altshuler, D.L., Auton, A., Brooks, L.D., Gibbs, R.A., Hurles, M.E., and McVean, G.A. (2010). A map of human genome variation from population-scale sequencing. Nature *467*, 1061–1073.

Fu, W., O'Connor, T.D., Jun, G., Kang, H.M., Abecasis, G., Leal, S.M., Gabriel, S., Altshuler, D., Shendure, J., Nickerson, D.A., *et al.* (2013). Analysis of 6,515 exomes reveals the recent origin of most human protein-coding variants. Nature *493*, 216–220.

Gibson, W.T., Hood, R.L., Zhan, S.H., Bulman, D.E., Fejes, A.P., Moore, R., Mungall, A.J., Eydoux, P., Babul-Hirji, R., An, J., *et al.* (2012). Mutations in EZH2 cause Weaver syndrome. Am. J. Hum. Genet. *90*, 110–118.

Gnirke, A., Melnikov, A., Maguire, J., Rogov, P., LeProust, E.M., Brockman, W., Fennell, T., Giannoukos, G., Fisher, S., Russ, C., *et al.* (2009). Solution hybrid selection with ultra-long oligonucleotides for massively parallel targeted sequencing. Nat. Biotechnol. *27*, 182–189.

Gui, Y., Guo, G., Huang, Y., Hu, X., Tang, A., Gao, S., Wu, R., Chen, C., Li, X., Zhou, L., *et al.* (2011). Frequent mutations of chromatin remodeling genes in transitional cell carcinoma of the bladder. Nat. Genet. *43*, 875–878.

Guo, B.R., Zhang, X., Chen, G., Zhang, J.G., Sun, L.D., Du, W.D., Zhang, Q., Cui, Y., Zhu, J., Tang, X.F., *et al.*

(2012a). Exome sequencing identifies a COL14A1 mutation in a large Chinese pedigree with punctate palmoplantar keratoderma. J. Med. Genet. *49*, 563–568.

Guo, G., Gui, Y., Gao, S., Tang, A., Hu, X., Huang, Y., Jia, W., Li, Z., He, M., Sun, L., *et al.* (2012b). Frequent mutations of genes encoding ubiquitin-mediated proteolysis pathway components in clear cell renal cell carcinoma. Nat. Genet. *44*, 17–19.

Hodges, E., Xuan, Z., Balija, V., Kramer, M., Molla, M.N., Smith, S.W., Middle, C.M., Rodesch, M.J., Albert, T.J., Hannon, G.J., *et al.* (2007). Genome-wide in situ exon capture for selective resequencing. Nat. Genet. *39*, 1522–1527.

Hoischen, A., van Bon, B.W., Gilissen, C., Arts, P., van Lier, B., Steehouwer, M., de Vries, P., de Reuver, R., Wieskamp, N., Mortier, G., *et al.* (2010). De novo mutations of SETBP1 cause Schinzel-Giedion syndrome. Nat. Genet. *42*, 483–485.

Hoischen, A., van Bon, B.W., Rodriguez-Santiago, B., Gilissen, C., Vissers, L.E., de Vries, P., Janssen, I., van Lier, B., Hastings, R., Smithson, S.F., *et al.* (2011). De novo nonsense mutations in ASXL1 cause Bohring-Opitz syndrome. Nat. Genet. *43*, 729–731.

Holm, H., Gudbjartsson, D.F., Sulem, P., Masson, G., Helgadottir, H.T., Zanon, C., Magnusson, O.T., Helgason, A., Saemundsdottir, J., Gylfason, A., *et al.* (2011). A rare variant in MYH6 is associated with high risk of sick sinus syndrome. Nat. Genet. *43*, 316–320.

Hou, Y., Song, L., Zhu, P., Zhang, B., Tao, Y., Xu, X., Li, F., Wu, K., Liang, J., Shao, D., *et al.* (2012). Single-cell exome sequencing and monoclonal evolution of a JAK2-negative myeloproliferative neoplasm. Cell *148*, 873–885.

Hudson, T.J., Anderson, W., Artez, A., Barker, A.D., Bell, C., Bernabe, R.R., Bhan, M.K., Calvo, F., Eerola, I., Gerhard, D.S., *et al.* (2010). International network of cancer genome projects. Nature *464*, 993–998.

Hunt, K.A., Mistry, V., Bockett, N.A., Ahmad, T., Ban, M., Barker, J.N., Barrett, J.C., Blackburn, H., Brand, O., Burren, O., *et al.* (2013). Negligible impact of rare autoimmune-locus coding-region variants on missing heritability. Nature *498*, 232–235.

Jiang, T., Yang, L., Jiang, H., Tian, G., and Zhang, X. (2011). High-performance single-chip exon capture allows accurate whole exome sequencing using the Illumina Genome Analyzer. Sci. China Life Sci. *54*, 945–952.

Jones, S., Wang, T.L., Shih Ie, M., Mao, T.L., Nakayama, K., Roden, R., Glas, R., Slamon, D., Diaz, L.A., Jr., Vogelstein, B., *et al.* (2010). Frequent mutations of chromatin remodeling gene ARID1A in ovarian clear cell carcinoma. Science *330*, 228–231.

Jonsson, T., Atwal, J.K., Steinberg, S., Snaedal, J., Jonsson, P.V., Bjornsson, S., Stefansson, H., Sulem, P., Gudbjartsson, D., Maloney, J., *et al.* (2012). A mutation in APP protects against Alzheimer's disease and age-related cognitive decline. Nature *488*, 96–99.

Lam, H.Y., Pan, C., Clark, M.J., Lacroute, P., Chen, R., Haraksingh, R., O'Huallachain, M., Gerstein, M.B., Kidd, J.M., Bustamante, C.D., *et al.* (2012). Detecting and annotating genetic variations using the HugeSeq pipeline. Nat. Biotechnol. *30*, 226–229.

Langmead, B., Schatz, M.C., Lin, J., Pop, M., and Salzberg, S.L. (2009). Searching for SNPs with cloud computing. Genome Biol. *10*, R134.

Levy, S., Sutton, G., Ng, P.C., Feuk, L., Halpern, A.L., Walenz, B.P., Axelrod, N., Huang, J., Kirkness, E.F., Denisov, G., *et al.* (2007). The diploid genome sequence of an individual human. PLoS Biol. *5*, e254.

Li, B., and Leal, S.M. (2008). Methods for detecting associations with rare variants for common diseases: application to analysis of sequence data. Am. J. Hum. Genet. *83*, 311–321.

Li, H., Handsaker, B., Wysoker, A., Fennell, T., Ruan, J., Homer, N., Marth, G., Abecasis, G., and Durbin, R. (2009a). The Sequence Alignment/Map format and SAMtools. Bioinformatics *25*, 2078–2079.

Li, R., Li, Y., Fang, X., Yang, H., Wang, J., and Kristiansen, K. (2009b). SNP detection for massively parallel whole-genome resequencing. Genome Res. *19*, 1124–1132.

Li, S., Li, R., Li, H., Lu, J., Li, Y., Bolund, L., Schierup, M.H., and Wang, J. (2013). SOAPindel: efficient identification of indels from short paired reads. Genome Res. *23*, 195–200.

Li, Y., Vinckenbosch, N., Tian, G., Huerta-Sanchez, E., Jiang, T., Jiang, H., Albrechtsen, A., Andersen, G., Cao, H., Korneliussen, T., *et al.* (2010). Resequencing of 200 human exomes identifies an excess of low-frequency non-synonymous coding variants. Nat. Genet. *42*, 969–972.

Li, Y., Sidore, C., Kang, H.M., Boehnke, M., and Abecasis, G.R. (2011). Low-coverage sequencing: implications for design of complex trait association studies. Genome Res. *21*, 940–951.

Lin, Z., Chen, Q., Lee, M., Cao, X., Zhang, J., Ma, D., Chen, L., Hu, X., Wang, H., Wang, X., *et al.* (2012). Exome sequencing reveals mutations in TRPV3 as a cause of Olmsted syndrome. Am. J. Hum. Genet. *90*, 558–564.

Liu, D.J., and Leal, S.M. (2010). A novel adaptive method for the analysis of next-generation sequencing data to detect complex trait associations with rare variants due to gene main effects and interactions. PLoS Genet. *6*, e1001156.

Liu, Y., Gao, M., Lv, Y.M., Yang, X., Ren, Y.Q., Jiang, T., Zhang, X., Guo, B.R., Li, M., Zhang, Q., *et al.* (2011). Confirmation by Exome Sequencing of the Pathogenic Role of NCSTN Mutations in Acne Inversa (Hidradenitis Suppurativa). J. Invest. Dermatol. *131*, 1570–1572.

Lupski, J.R., Belmont, J.W., Boerwinkle, E., and Gibbs, R.A. (2011). Clan genomics and the complex architecture of human disease. Cell *147*, 32–43.

Lyon, G.J., Jiang, T., Van Wijk, R., Wang, W., Bodily, P.M., Xing, J., Tian, L., Robison, R.J., Clement, M., Lin, Y., *et al.* (2011). Exome sequencing and unrelated findings in the context of complex disease research: ethical and clinical implications. Discov. Med. *12*, 41–55.

Madsen, B.E., and Browning, S.R. (2009). A groupwise association test for rare mutations using a weighted sum statistic. PLoS Genet. *5*, e1000384.

Mamanova, L., Coffey, A.J., Scott, C.E., Kozarewa, I., Turner, E.H., Kumar, A., Howard, E., Shendure, J., and Turner, D.J. (2010). Target-enrichment strategies for next-generation sequencing. Nat. Methods *7*, 111–118.

Manolio, T.A., Collins, F.S., Cox, N.J., Goldstein, D.B., Hindorff, L.A., Hunter, D.J., McCarthy, M.I., Ramos, E.M., Cardon, L.R., Chakravarti, A., *et al.* (2009). Finding the missing heritability of complex diseases. Nature *461*, 747–753.

McClellan, J., and King, M.C. (2010). Genetic heterogeneity in human disease. Cell *141*, 210–217.

Mathieson, I., and McVean, G. (2012). Differential confounding of rare and common variants in spatially structured populations. Nat. Genet. *44*, 243–246.

Maurano, M.T., Humbert, R., Rynes, E., Thurman, R.E., Haugen, E., Wang, H., Reynolds, A.P., Sandstrom, R., Qu, H., Brody, J., *et al.* (2012). Systematic localization of common disease-associated variation in regulatory DNA. Science *337*, 1190–1195.

Musunuru, K., Pirruccello, J.P., Do, R., Peloso, G.M., Guiducci, C., Sougnez, C., Garimella, K.V., Fisher, S., Abreu, J., Barry, A.J., *et al.* (2010). Exome sequencing, ANGPTL3 mutations, and familial combined hypolipidemia. N. Engl. J. Med. *363*, 2220–2227.

Neale, B.M., Rivas, M.A., Voight, B.F., Altshuler, D., Devlin, B., Orho-Melander, M., Kathiresan, S., Purcell, S.M., Roeder, K., and Daly, M.J. (2011). Testing for an unusual distribution of rare variants. PLoS Genet. *7*, e1001322.

Nelson, M.R., Wegmann, D., Ehm, M.G., Kessner, D., St Jean, P., Verzilli, C., Shen, J., Tang, Z., Bacanu, S.A., Fraser, D., *et al.* (2012). An abundance of rare functional variants in 202 drug target genes sequenced in 14,002 people. Science *337*, 100–104.

Ng, P.C., and Henikoff, S. (2003). SIFT: Predicting amino acid changes that affect protein function. Nucleic Acids Res. *31*, 3812–3814.

Ng, P.C., Levy, S., Huang, J., Stockwell, T.B., Walenz, B.P., Li, K., Axelrod, N., Busam, D.A., Strausberg, R.L., and Venter, J.C. (2008). Genetic variation in an individual human exome. PLoS Genet. *4*, e1000160.

Ng, S.B., Turner, E.H., Robertson, P.D., Flygare, S.D., Bigham, A.W., Lee, C., Shaffer, T., Wong, M., Bhattacharjee, A., Eichler, E.E., *et al.* (2009). Targeted capture and massively parallel sequencing of 12 human exomes. Nature *461*, 272–276.

Ng, S.B., Buckingham, K.J., Lee, C., Bigham, A.W., Tabor, H.K., Dent, K.M., Huff, C.D., Shannon, P.T., Jabs, E.W., Nickerson, D.A., *et al.* (2010). Exome sequencing identifies the cause of a mendelian disorder. Nat. Genet. *42*, 30–35.

Nielsen, R., Paul, J.S., Albrechtsen, A., and Song, Y.S. (2011). Genotype and SNP calling from next-generation sequencing data. Nat. Rev. Genet. *12*, 443–451.

O'Rawe, J., Jiang, T., Sun, G., Wu, Y., Wang, W., Hu, J., Bodily, P., Tian, L., Hakonarson, H., Johnson, W.E., *et al.* (2013). Low concordance of multiple variant-calling pipelines: practical implications for exome and genome sequencing. Genome Med. *5*, 28.

O'Roak, B.J., Deriziotis, P., Lee, C., Vives, L., Schwartz, J.J., Girirajan, S., Karakoc, E., Mackenzie, A.P., Ng, S.B., Baker, C., *et al.* (2011). Exome sequencing in sporadic autism spectrum disorders identifies severe de novo mutations. Nat. Genet. *43*, 585–589.

Pasaniuc, B., Rohland, N., McLaren, P.J., Garimella, K., Zaitlen, N., Li, H., Gupta, N., Neale, B.M., Daly, M.J., Sklar, P., *et al.* (2012). Extremely low-coverage sequencing and imputation increases power for genome-wide association studies. Nat. Genet. *44*, 631–635.

Pasqualucci, L., Trifonov, V., Fabbri, G., Ma, J., Rossi, D., Chiarenza, A., Wells, V.A., Grunn, A., Messina, M., Elliot, O., *et al.* (2011). Analysis of the coding genome of diffuse large B-cell lymphoma. Nat. Genet. *43*, 830–837.

Pearson, T.A., and Manolio, T.A. (2008). How to interpret a genome-wide association study. JAMA *299*, 1335–1344.

Pleasance, E.D., Cheetham, R.K., Stephens, P.J., McBride, D.J., Humphray, S.J., Greenman, C.D., Varela, I., Lin, M.L., Ordonez, G.R., Bignell, G.R., *et al.* (2010). A comprehensive catalogue of somatic mutations from a human cancer genome. Nature *463*, 191–196.

Pollard, K.S., Hubisz, M.J., Rosenbloom, K.R., and Siepel, A. (2010). Detection of nonneutral substitution rates on mammalian phylogenies. Genome Res. *20*, 110–121.

Price, A.L., Kryukov, G.V., de Bakker, P.I., Purcell, S.M., Staples, J., Wei, L.J., and Sunyaev, S.R. (2010). Pooled association tests for rare variants in exon-resequencing studies. Am. J. Hum. Genet. *86*, 832–838.

Quadri, M., Fang, M., Picillo, M., Olgiati, S., Breedveld, G.J., Graafland, J., Wu, B., Xu, F., Erro, R., Amboni, M., *et al.* (2013). Mutation in the SYNJ1 gene associated with autosomal recessive, early-onset Parkinsonism. Hum. Mutat. *34*, 1208–1215.

Reumers, J., De Rijk, P., Zhao, H., Liekens, A., Smeets, D., Cleary, J., Van Loo, P., Van Den Bossche, M., Catthoor, K., Sabbe, B., *et al.* (2012). Optimized filtering reduces the error rate in detecting genomic variants by short-read sequencing. Nat. Biotechnol. *30*, 61–68.

Rivas, M.A., Beaudoin, M., Gardet, A., Stevens, C., Sharma, Y., Zhang, C.K., Boucher, G., Ripke, S., Ellinghaus, D., Burtt, N., *et al.* (2011). Deep resequencing of GWAS loci identifies independent rare variants associated with inflammatory bowel disease. Nat. Genet. *43*, 1066–1073.

Rope, A.F., Wang, K., Evjenth, R., Xing, J., Johnston, J.J., Swensen, J.J., Johnson, W.E., Moore, B., Huff, C.D., Bird, L.M., *et al.* (2011). Using VAAST to identify an X-linked disorder resulting in lethality in male infants due to N-terminal acetyltransferase deficiency. Am. J. Hum. Genet. *89*, 28–43.

Sherry, S.T., Ward, M.H., Kholodov, M., Baker, J., Phan, L., Smigielski, E.M., and Sirotkin, K. (2001). dbSNP: the NCBI database of genetic variation. Nucleic Acids Res. *29*, 308–311.

Shi, Y., Li, Y., Zhang, D., Zhang, H., Lu, F., Liu, X., He, F., Gong, B., Cai, L., Li, R., *et al.* (2011). Exome sequencing identifies ZNF644 mutations in high myopia. PLoS Genet. *7*, e1002084.

Sjoblom, T., Jones, S., Wood, L.D., Parsons, D.W., Lin, J., Barber, T.D., Mandelker, D., Leary, R.J., Ptak, J., Silliman, N., *et al.* (2006). The consensus coding sequences of human breast and colorectal cancers. Science *314*, 268–274.

Stitziel, N.O., Kiezun, A., and Sunyaev, S. (2011). Computational and statistical approaches to analyzing variants identified by exome sequencing. Genome Biol. *12*, 227.

Stratton, M.R., Campbell, P.J., and Futreal, P.A. (2009). The cancer genome. Nature *458*, 719–724.

Sulem, P., Gudbjartsson, D.F., Walters, G.B., Helgadottir, H.T., Helgason, A., Gudjonsson, S.A., Zanon, C., Besenbacher, S., Bjornsdottir, G., Magnusson, O.T., *et al.* (2011). Identification of low-frequency variants associated with gout and serum uric acid levels. Nat. Genet. *43*, 1127–1130.

Tennessen, J.A., Bigham, A.W., O'Connor, T.D., Fu, W., Kenny, E.E., Gravel, S., McGee, S., Do, R., Liu, X., Jun, G., *et al.* (2012). Evolution and functional impact of rare coding variation from deep sequencing of human exomes. Science *337*, 64–69.

Vissers, L.E., de Ligt, J., Gilissen, C., Janssen, I., Steehouwer, M., de Vries, P., van Lier, B., Arts, P., Wieskamp, N., del Rosario, M., *et al.* (2010). A de novo paradigm for mental retardation. Nat. Genet. *42*, 1109–1112.

Wang, J.L., Yang, X., Xia, K., Hu, Z.M., Weng, L., Jin, X., Jiang, H., Zhang, P., Shen, L., Guo, J.F., *et al.* (2010). TGM6 identified as a novel causative gene of spinocerebellar ataxias using exome sequencing. Brain *133*, 3510–3518.

Wang, J.L., Cao, L., Li, X.H., Hu, Z.M., Li, J.D., Zhang, J.G., Liang, Y., San, A., Li, N., Chen, S.Q., *et al.* (2011). Identification of PRRT2 as the causative gene of paroxysmal kinesigenic dyskinesias. Brain *134*, 3493–3501.

Wang, K., Li, M., and Hakonarson, H. (2010). ANNOVAR: functional annotation of genetic variants from high-throughput sequencing data. Nucleic Acids Res. *38*, e164.

Wei, Z., Wang, W., Hu, P., Lyon, G.J., and Hakonarson, H. (2011). SNVer: a statistical tool for variant calling in analysis of pooled or individual next-generation sequencing data. Nucleic Acids Res. *39*, e132.

Xu, X., Hou, Y., Yin, X., Bao, L., Tang, A., Song, L., Li, F., Tsang, S., Wu, K., Wu, H., *et al.* (2012). Single-cell exome sequencing reveals single-nucleotide mutation characteristics of a kidney tumor. Cell *148*, 886–895.

Yan, X.J., Xu, J., Gu, Z.H., Pan, C.M., Lu, G., Shen, Y., Shi, J.Y., Zhu, Y.M., Tang, L., Zhang, X.W., *et al.* (2011). Exome sequencing identifies somatic mutations of DNA methyltransferase gene DNMT3A in acute monocytic leukemia. Nat. Genet. *43*, 309–315.

Yandell, M., Huff, C., Hu, H., Singleton, M., Moore, B., Xing, J., Jorde, L.B., and Reese, M.G. (2011). A probabilistic disease-gene finder for personal genomes. Genome Res. *21*, 1529–1542.

Yi, X., Liang, Y., Huerta-Sanchez, E., Jin, X., Cuo, Z.X., Pool, J.E., Xu, X., Jiang, H., Vinckenbosch, N., Korneliussen, T.S., *et al.* (2010). Sequencing of 50 human exomes reveals adaptation to high altitude. Science *329*, 75–78.

Zhang, C., Li, D., Zhang, J., Chen, X., Huang, M., Archacki, S., Tian, Y., Ren, W., Mei, A., Zhang, Q., *et al.* (2013). Mutations in ABCB6 Cause Dyschromatosis Universalis Hereditaria. J. Invest. Dermatol. *133*, 2221–2228.

Zhang, S.Q., Jiang, T., Li, M., Zhang, X., Ren, Y.Q., Wei, S.C., Sun, L.D., Cheng, H., Li, Y., Yin, X.Y., *et al.* (2012a). Exome sequencing identifies MVK mutations in disseminated superficial actinic porokeratosis. Nat. Genet. *44*, 1156–1160.

Zhang, X., Guo, B.R., Cai, L.Q., Jiang, T., Sun, L.D., Cui, Y., Hu, J.C., Zhu, J., Chen, G., Tang, X.F., *et al.* (2012b). Exome sequencing identified a missense mutation of EPS8L3 in Marie Unna hereditary hypotrichosis. J. Med. Genet. *49*, 727–730.

Insights on Plant Development Using NGS Technologies

8

Ying Wang and Yuling Jiao

Abstract

Despite being a recent breakthrough technology, Next-generation sequencing (NGS) has already found its application in plant developmental biology. This review summarizes recent applications of NGS in understanding key questions in plant development, such as organ formation and genetic information transmission in germline cells. NGS has been used to study gene expression, transcription factor (TF) binding, and epigenetic regulation, which includes DNA methylation, histone modification, and profiles of small RNA (smRNA) expression. NGS has been combined with traditional and novel technologies to address these types of questions that have intrigued plant developmental biologists. This review will cover recent advances in cell-specific analysis of transcription by NGS, and their applications to understand organ formation and development in two model species *Arabidopsis thaliana* and maize (*Zea mays*).

Introduction

NGS has revolutionized biological research including plant development, in which scientists study the genetic control of cell growth, differentiation and morphogenesis, the process that gives rise to tissues, organs, other anatomical features, and their spatial organizations. Studies of plant development at the molecular level, mostly by molecular genetic approaches, have identified numerous transcription factors influencing plant development, indicating complex transcriptional regulatory networks underlying each developmental process (Kaufmann *et al.*, 2010a; Wellmer and Riechmann, 2010). NGS has been applied in plant developmental biology research to explore at least three types of regulatory mechanisms, genome-wide changes of transcription, the association of TFs and their cognate binding sites, and epigenetic modifications during development. For the first type of questions, DNA microarrays have traditionally been utilized to study transcriptomes (Fodor *et al.*, 1993; Schena *et al.*, 1995). More recently, NGS-based RNA-seq has become an alternative to microarray with better accuracy, resolution and coverage (Mortazavi *et al.*, 2008; Nagalakshmi *et al.*, 2008; Wilhelm *et al.*, 2008). RNA-seq uses NGS to sequence cDNA in order to get information about RNA content in a biological sample, and is therefore also called 'whole transcriptome shotgun sequencing'. The last two types of questions require technologies such as chromatin immunoprecipitation (ChIP), which will be further elaborated below. The immunoprecipitated DNA-associated protein can be TFs or other chromatin binding proteins. Although DNA tiling microarrays covering both coding regions and non-coding regions were initially used to profile target protein binding sequences (ChIP-chip) (Iyer *et al.*, 2001; Ren *et al.*, 2000), NGS approaches were soon found superior to tiling microarrays due to NGS's high resolution, low background, reduced demand of sample input, and even more affordable cost (Barski *et al.*, 2007; Robertson *et al.*, 2007). In such applications, which are usually termed ChIP-seq, immunoprecipitated DNA fragments are amplified and directly sequenced by NGS platforms.

In this review, we summarize recent applications of NGS in plant developmental biology by reviewing the use of RNA-seq and ChIP-seq to address developmental questions.

Use RNA-seq to dissect transcription at the cellular resolution

The acquisition of distinct cell fates is central to the development of multicellular organisms and is largely mediated by gene expression patterns specific to individual cells and tissues. An analysis of gene expression with spatial and temporal resolution facilitates the elucidation of transcriptional networks linked to cellular identity and function. NGS-based RNA-seq has been used to study transcriptome changes during development, to identify genes affected in developmental mutants at the genome scale, and to identify genes induced or repressed after the induction of key developmental regulators. Among the different approaches, cell-type specific transcriptome studies represent a typical use of RNA-seq to study development. Not until recently, has such a cellular level of resolution been available for a comprehensive view of the entire transcriptome afforded by profiling methods (Wang and Jiao, 2011). Indeed, NGS has quickly been combined with approaches for cell-specific transcriptome profiling to extend transcriptome analysis to the single-base resolution.

Because of its importance in developmental studies, especially the recent new progress in this field, cell-type specific gene expression profiling will be the focus of this review. Usually, the separation of distinct cell types implies physical manipulations on the basis of their distinct appearance, location, molecular markers, or other distinguishable properties among different cell types (Nelson *et al.*, 2008). Although the separation of certain cell types (such as pollen, trichomes, and guard cells) based on their unique properties has long been available, new technologies are required for the profiling of most other cell types in plants.

Laser microdissection (LM) is a method for separating and isolating specific cells of interest from microscopic regions of tissues (Emmert-Buck *et al.*, 1996). LM does not require cell-specific markers or tags, unique genetic lines or special tissue or cell properties; cells are selected entirely on the basis of their appearance and location in histological sections. During LM isolation, 5–10 µm tissue sections are visualized under a microscope, and cells of interest are identified and isolated.

The limited number of cells that can be harvested by LM requires efficient RNA isolation at the femtogram to picogram scale as well as RNA amplification to enable RNA-seq. Traditionally RNA amplification is achieved by *in vitro* transcription (IVT), which is based on two to three rounds of antisense RNA (aRNA) amplification (Van Gelder *et al.*, 1990), and involves a series of enzymatic reactions resulting in amplification of exceedingly small amounts of RNA obtained by LM. Recently, an alternative approach has been proposed and used in animal and plant development studies. This new strategy utilizes a terminal transferase to add poly(A) tails to newly synthesized first-strand cDNA. The additional poly(A) tail together with the poly(A) tail at the 3′ end of mRNAs enables subsequent PCR amplification (Kurimoto *et al.*, 2006). It was shown that this new approach has higher sensitivity and lower amplification bias than IVT (Tang *et al.*, 2009).

Both these amplification approaches have been combined with NGS library preparation protocols for transcriptome profiling to study plant development. By isolating leaf bundle sheath and mesophyll cells from four different developing zones along the distal-proximal axes of maize leaves by LM followed by IVT amplification and NGS, Li and colleagues discovered a developmental gradient along the distal-proximal axes (Li *et al.*, 2010). Over half of the genome shows differential expression along the developmental gradient, including many key functional groups such as cell wall biosynthesis, chloroplast targeting, and photosynthesis. A 21% difference was identified between bundle sheath and mesophyll cells, which may drive C_4 photosynthetic differentiation in the maize leaf blade. More recently, the poly(A)-adding based amplification method was combined with LM to profile transcription network in *Arabidopsis* female gametophyte, one of the female gametes harboured in the reproductive organs of the flower (Schmid *et al.*, 2012). The NGS-based approach identified more than twice the number of expressed genes compared with previous reports using LM followed by IVT amplification and Affymetrix microarray hybridization-based detection (Wuest *et al.*, 2010). In addition, several classes of genes that were systematically underrepresented by the microarray studies were identified by NGS, including many genes with potential important roles in developmental processes and specific cellular functions. All these findings highlighted the importance of using NGS to study transcriptome in order to understand plant development.

Whereas LM harvests total RNA from isolated cells, a new approach, translating ribosome affinity purification (TRAP), immunoprecipitates labelled polysomes directly from tissue homogenates obtained from organisms expressing an epitope-tag labelled ribosomal protein. In *Arabidopsis*, these plants express an FLAG-tagged 60S ribosomal protein L18 (RPL18) under cell-specific promoters to restrict its expression in target cells (Jiao and Meyerowitz, 2010; Mustroph *et al.*, 2009). By immunoprecipitating FLAG-tagged ribosomes, TRAP detects only ribosome-associated mRNAs, and therefore only transcripts that are actively being translated, rather than the full population of transcribed RNA, making it a better proxy to protein synthesis of living cells (Zanetti *et al.*, 2005). More importantly, cell-specific expression of labelled RPL18 adds cell type-specificity to TRAP by utilizing cell type-specific labelling and isolating translating transcripts for genome-wide profiling. TRAP usually achieves high RNA yield to eliminate information-losing, bias-introducing, and high-cost RNA amplification, making it possible to directly combine TRAP with the NGS based RNA-seq technology (Jiao and Meyerowitz, 2010). Different from LM or FACS, TRAP does not require expensive and specialized equipment that is not generally available and is thus more widely affordable.

The combination of TRAP and RNA-seq (TRAP-seq) has recently been used to study early flower development in *Arabidopsis* (Jiao and Meyerowitz, 2010). Flower development is one of the best-understood aspects of plant development. Although the organ formation process is as yet poorly understood, we have a relatively clear understanding of organ identity specification (for review, see Krizek and Fletcher, 2005). Genetic studies identified floral organ identity genes and proposed the ABC model, in which three classes of regulatory genes, A, B and C, work in a combinatorial manner to confer organ identities of four floral whorls (Coen and Meyerowitz, 1991). Each class of regulatory genes is expressed in a specific and evolutionarily conserved domain, and the action of the class A, B and C genes is necessary for specification of organ identity. Using TRAP-seq, cell type-specific mRNAs were isolated from the A, B or C domain, and quantified by NGS. To achieve temporal specificity, a floral induction system was further used to facilitate collection of early stage flowers (Wellmer *et al.*, 2006). In this genetically engineered floral induction system, the expression of *APETALA1* (*AP1*) is under the control of dexamethasone to synchronize flower initiation upon chemical treatment. The integration of domain-specific promoters and this floral induction system resulted in fine-scale spatiotemporal expression of HF–RPL18, which allowed the isolation of translating mRNA in specific cellular domains, and at specific developmental stages. Comparisons among the three cell types not only identified known cell-specific transcripts but also uncovered many new ones, from which cell type-specific hormone responses were inferred, *cis*-acting promoter motifs and co-expressed cognate binding factor candidates were identified. Several dozens of cell type-specific splicing isoforms were also identified through single-base resolution NGS. As only translating mRNAs were isolated, these splicing isoforms did not include splicing defects without translation. By comparing translating mRNAs from TRAP with steady-state overall transcripts, evidence for widespread post-transcriptional regulation at both the intron splicing and translational stages were found. Sequence analyses of post-transcriptionally regulated transcripts further revealed the existence of several types of structural features associated with each regulatory step.

There are two other approaches for cell type-specific profiling, and in principle, both can be combined with NGS. Fluorescence-activated cell sorting (FACS) employs flow cytometry, a routine technique for counting and examining blood cells through an electronic detection apparatus

(Birnbaum *et al.*, 2003). The adoption of FACS in plants needs to overcome a unique challenge in that single cells are difficult to obtain as plant organs and tissues are complex three-dimensional assemblies with each cell surrounded by cell walls. To address this problem, protoplasts of several cell types have been isolated after enzymatic digestion of cell walls to facilitate cell sorting of fluorescent protein-labelled cell types. An alternative to whole protoplast isolation is nuclear labelling and isolation. Nuclear isolation and sorting are significantly easier and faster than cell protoplasting and sorting (Zhang *et al.*, 2008). Two variants of nuclei sorting are currently available. The first approach is similar to FACS except that the GFP or other fluorescent protein marker must be targeted to the nuclei of interest so that they can be isolated and sorted (Zhang *et al.*, 2008). The second approach, termed INTACT for Isolation of Nuclei TAgged in specific Cell Types, is somewhat similar to TRAP in that nuclei are biotin-labelled through a nuclear envelope protein for affinity isolation (Deal and Henikoff, 2010). Total nuclei are isolated from transgenic plants and biotin-labelled nuclei are then purified using streptavidin-coated magnetic beads. Comparably high yield and purity can be achieved by INTACT as in TRAP (Deal and Henikoff, 2010; Jiao and Meyerowitz, 2010). Although both methods for nuclei isolation have one potential drawback that RNA profiles of nuclei are clearly different from cytoplasmic mRNA profiles, and therefore not representative of mRNAs being translated into proteins, the isolated nuclei can be used to profile chromatin features in addition to gene expression (Deal and Henikoff, 2010).

Use ChIP-seq to dissect transcriptional networks

Protein–DNA interactions are central to development, both in normal chromatin assembly and in the binding of *cis*-elements by TFs. It is widely assumed that *cis*-elements of genomic DNA integrate cellular signals by the tethering of TFs, chromatin associated proteins, and the basal transcription machinery. Understanding the binding sites and recognition *cis*-elements of TFs that control plant development poses challenges in understanding transcriptional regulatory networks underlying plant development.

Several experimental and computational methods have been developed to identify protein–DNA interactions. Electrophoretic mobility shift assays (EMSA) and chemical or nuclease footprinting are widely used to explore the specific interaction of a given fragment of DNA and a TF. These techniques, however, are time-consuming and involve *in vitro* protein–DNA interactions. By contrast, chromatin immunoprecipitation (ChIP) provides a versatile and powerful tool to investigate protein–DNA interactions *in vivo*. ChIP experiment starts with intact tissues or cells, in which cross-linking reagent, such as formaldehyde, covalently cross-links a protein to its bound DNA (Gilmour and Lis, 1984). The chromatin is then extracted and sheared, and the protein–DNA complex is immunoprecipitated by specific antibodies against the DNA-binding protein. For specific genomic regions of interest, the enrichment of protein-binding DNA can be quantified by real-time PCR. More generally, the enriched protein-binding DNA can be analysed by hybridization to genome-tiling microarray (ChIP-chip) or by direct NGS (ChIP-seq) to identify all binding sites in the genome.

Although ChIP-chip was developed earlier, NGS-based ChIP-seq has many advantages over ChIP-chip. First of all, ChIP-chip requires whole genome microarray, which is costly, and not available for many species. In addition, ChIP-chip requires larger amount of input DNA sample, which are usually extensively amplified and tend to introduce systematic bias. Finally, microarray has limited sensitivity, specificity, and dynamic range compared with NGS data. Thus, compared to ChIP-chip, ChIP-seq has become the method of choice in most applications in plant development studies.

ChIP-seq has been used in an expanding number of studies on plant development, including meristem organization, flower development, and photomorphogenesis (Bolduc *et al.*, 2012; Busch *et al.*, 2010; Deng *et al.*, 2011; Kaufmann *et al.*, 2009, 2010b; Ouyang *et al.*, 2011; Yant *et al.*, 2010). Owing to space limitations, this review will focus on the application of ChIP-seq in flower development (Fig. 8.1).

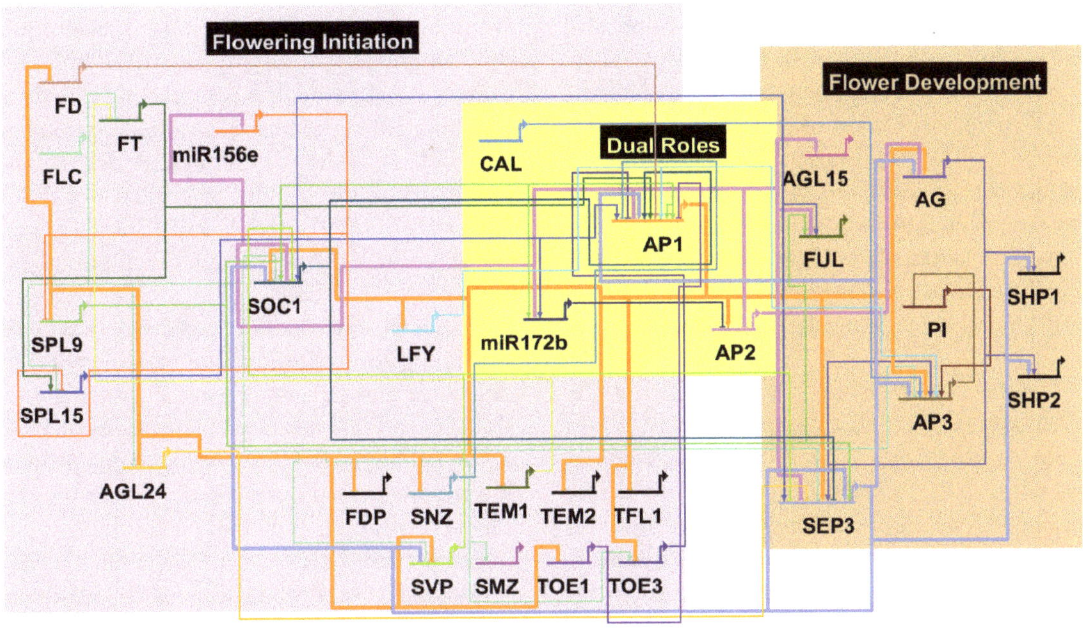

Figure 8.1 Summary of regulatory networks controlling *Arabidopsis* flower development. Direct interactions identified by NGS-based ChIP-seq are highlighted by bold lines (Kaufmann *et al.*, 2009, 2010b; Yant *et al.*, 2010). The network was visualized using the BioTapestry program.

Flower development is a classic model for developmental biology (Krizek and Fletcher, 2005). In the above-mentioned ABC model for flower development, floral organs are specified by the combinatorial activity of three functional gene classes. The A function genes in the model plant *Arabidopsis* are represented by *AP1* and *AP2* specifying sepal identity. The A function genes determine petal identity together with the B function genes *AP3* and *PISTILLATA* (*PI*). The C function gene *AGAMOUS* (*AG*) alone determines carpel identity and, together with the B function genes, specifies stamen identity. Additional E function genes, *SEPALLATA1–4* (*SEP1–4*), are required for the specification of all four types of floral organs (Pelaz *et al.*, 2000). Based on genetic and biochemical results, it was proposed that the combinatorial protein interactions of those ABCE function MADS box TFs assemble organ-specific quaternary protein complexes to specify floral organs (Honma and Goto, 2001; Riechmann *et al.*, 1996a; Smaczniak *et al.*, 2012). It has been demonstrated that such protein complexes can bind to two CArG boxes with a consensus sequence CC[A/T]$_6$GG within the regulatory regions of target gene promoters (Huang *et al.*, 1993; Riechmann *et al.*, 1996b; Shiraishi *et al.*, 1993). More recently, NGS-based ChIP-seq results revealed the potential global regulatory networks downstream of MADS protein complexes that control floral organ identity switches.

SEP3, as well as three related E function SEP proteins, interacts with other MADS-domain transcription factors to mediate the formation of higher-order TF complexes (Honma and Goto, 2001; Smaczniak *et al.*, 2012). Through ChIP-seq and ChIP-chip experiments, it was shown that SEP3 binds to thousands of regions throughout the genome to regulate potential target genes (Kaufmann *et al.*, 2009). Among its targets, many also encode TFs. Genome-wide mapping of binding sites confirmed previously identified binding motif CArG boxes by *in vitro*-binding assays. In addition, the NGS results confirmed the predicted existence of a nearby second CArG box corresponding to the formation of multimeric MADS protein complexes that bind to a nearby CArG box. In addition to the CArG box, SEP3-binding regions were also found enriched with the auxin-responsive AuxRE and light-responsive G-box. These findings imply a possible crosstalk between flower development and other signalling pathways, such as auxin response. Direct interactions were identified between SEP3 and promoters of most other MADS-box TFs controlling flower development, including SEP3 itself and the other three SEP genes. These direct interactions likely promote target gene expression, consistent with SEP3 being a master regulator of flower development.

Another MADS-box TF investigated by ChIP-seq was AP1 (Kaufmann *et al.*, 2010b). In addition to being an A function gene, AP1 and its paralogue CAULIFLOWER (CAL) redundantly promote flowering, primarily the onset of flower development (Irish and Sussex, 1990). Indeed, AP1 is initially expressed throughout emerging floral primordial, and later restricted to the outer two whorls of floral buds to specify sepals and petals as an A function gene (Mandel *et al.*, 1992). ChIP-seq identified approximately 2000 AP1 binding sites in the genome during early stages of floral meristem initiation. It was found that AP1 acts mostly as a transcriptional repressor to suppress previous developmental program for meristem functions during floral meristem initiation. In addition, AP1 down-regulates its own activators FD, SQUAMOSA PROMOTER BINDING PROTEIN-LIKE 9 (SPL9), and SUPPRESSOR OF OVEREXPRESSION OF CONSTANS 1 (SOC1). On the other hand, it up-regulates another activator of itself, LEAFY, which acts together with AP1 in floral meristem identity specification. After floral meristem initiation, AP1 functions mostly as a transcriptional activator to specify organ identity during early flower development. Notably, AP1 shares many of its targets with its dimerization partner SEP3.

ChIP-seq has also been used to map direct targets of another A function gene AP2 (Yant *et al.*, 2010). Different from all other ABCE function genes, which encode related MADS-box TFs, AP2 is the founding member of the plant-specific AP2-family TF (Jofuku *et al.*, 1994). Similar to AP1, AP2 has dual roles during early flower development (Bowman *et al.*, 1989). AP2 acts as a repressor of floral transition, and it promotes sepal and petal identity specification to function as another A gene later during organ specification. By identifying around 2000 genomic AP2-binding regions, it was found that AP2 represses the floral transition by direct transcriptional repression of many TFs or indirectly through binding to and activating microRNAs and other repressor genes. For example, AP2 activates miR156e to repress the expression of the flowering activating gene SPL.

Although the current regulatory networks underlying flower development remain to be fully elucidated and refined, the initial application of NGS-based ChIP-seq has largely expanded our understanding of this key developmental process, and provided numerous direct interactions among TFs (Fig. 8.1). Such genome-wide datasets can also be used for computational modelling of plant development (Alvarez-Buylla *et al.*, 2003; La Rota *et al.*, 2011; van Mourik *et al.*, 2010).

Use ChIP-seq to analyse the epigenome

The epigenetic regulation of chromatin structure is essential for the interpretation of genetic information and ultimately for the understanding of plant development. In fact, gene expression during plant development is influenced by genome-wide combinatorial interactions of epigenetic modifications (He *et al.*, 2011). NGS has led to high-resolution maps of plant epigenomes when combined with chromatin-analysis technologies.

Plant chromatin structure and gene expression are regulated by several epigenetic mechanisms, including DNA methylation, histone modifications, and part of small-interfering RNA (siRNA) pathways. DNA methylation refers to the formation of 5-methylcytosine by the addition of a methyl group to a cytosine base. DNA methylation regulates gene expression and suppresses the activity of transposons. Acetylation and methylation of lysine residues are two of the most studied histone modification marks. The quantitative combination of active and repressive histone modifications dynamically regulates genome accessibility and subsequently gene expression. siRNA are small RNAs (smRNAs) generated from double-stranded long precursor RNAs. Among siRNA subfamilies, 24 nt repeat-associated siRNA are derived from repeats and transposable elements. DNA methylation level is usually associated with 24 nt siRNA clusters.

Similar to ChIP study of TFs, NGS was initially used as an alternative to tiling microarray and has quickly become the method of choice to analyse epigenome regulation. To explore each type of epigenetic regulation, different chromatin technologies need to be combined with NGS. For DNA methylation analysis, bisulphite sequencing is the most used approach, and considered the gold standard for determining the methylation state of cytosines within a DNA sequence (Lister *et al.*, 2008). Sodium bisulphate treatment of genomic DNA under denaturing conditions converts cytosines to uracils yet leaves 5-methylcytosines unchanged (Frommer *et al.*, 1992), allowing quantification of cytosine methylation by NGS. Histone modification profiling relies on the development of highly specific histone antibodies that specifically interacts with modified histones. ChIP-seq with such antibodies has been used to determine the genomic locations of specific histone modifications. Genome-wide profiling of smRNA involves direct sequencing by NGS. Massively parallel signature sequencing (MPSS) (Brenner *et al.*, 2000), a prototype NGS, was initially used to analyse small RNAs (Lu *et al.*, 2005), before other NGS platforms were used for smRNA profiling. Briefly, smRNAs are purified by size selection, tag labelled, reverse transcribed, and amplified by PCR for direct sequencing. NGS-based profiling of the epigenome has been used to analyse plant development, including pollen development, endosperm development, and light regulation (Charron *et al.*, 2009; Gehring *et al.*, 2009; He *et al.*, 2010; Hsieh *et al.*, 2009; Wang *et al.*, 2009; Zhou *et al.*, 2010). Owing to space limitation, this review will focus on findings from pollen development, which is under extensive epigenetic regulation.

Pollen is the paternal germline cell that transmits genetic information to the next-generation. A high demand of genome stability maintenance is a salient feature of germline cells, including pollen (Henderson and Jacobsen, 2007). A recent study in *Arabidopsis* used NGS to analyse smRNA and DNA methylation in pollen, sperm cells, and non-inherited supporting cells (Slotkin *et al.*, 2009). As was found in other higher plants (Jiao and Deng, 2007), transposable elements (TEs) have limited but clear expression in vegetative cells, which are not inherited to the next-generation. It was found that some TEs are reactivated and transpose in the pollen vegetative nucleus, which accompanies the sperm cells but does not transmit DNA to the fertilized zygote. The mutagenic activity of TE is usually suppressed by epigenetic silencing at DNA methylation level through siRNAs. Consistently, TE methylation state is high in germline cells. Slotkin and colleagues used NGS to show a dramatic gain of 21 nt siRNAs generated from *Athila* retrotransposons in vegetative cells, whereas most 24 nt siRNAs, which are involved in RNA-direct DNA methylation, are generally low in these cells. The high expression *Athila* TEs in vegetative cells resulted in the movement of the derived 21 nt siRNAs into non-*Athila*-expressing sperm cells. Based on NGS results and other experiments, it has been proposed that transient TE reactivation in vegetative cells can induce TE silencing in the neighbouring sperm cells by transporting *Athila*-derived 21 nt siRNAs. This mechanism ensures genome stability in sperm cells by sacrificing the genome integrity of their supporting vegetative cells.

Conclusions and perspectives

Recent advances in NGS have enabled us to dissect transcriptional regulation and epigenetic regulation of plant development at the genome scale with unprecedented resolution. When combined with other molecular biology technologies, NGS has been used to study the transcriptome at cellular resolution, to identify direct binding regions of TFs, to analyse chromatin modifications, and to profile smRNAs. In plant development, all these regulatory mechanisms are involved. Therefore the use of NGS to study plant development has been very fruitful. With further development of sample preparation technologies, the reduction of sequencing cost, and the development of bioinformatics tools, we expect broader application of NGS in plant development studies in the near future. For example, the epigenome is dynamic and varies between cell types and responds to changes during development. Therefore, similar to transcriptome analysis, it is necessary to profile epigenomes at cellular resolution to understand the contributions of epigenomes to plant development from a dynamic viewpoint. New breakthroughs to label cell types, when combined with NGS, have the potential to lead our understanding of plant development to a higher resolution and at the genome scale.

Acknowledgements

The authors' laboratory is supported by the National Natural Science Foundation of China Grants 31171159 and 31222033, and by the National Basic Research Program of China (973 Program) Grant 2012CB910902. We wish to apologize to all authors whose works we could not cite due to space constraints.

References

Alvarez-Buylla, R., Huberman, A., Montero, S., Lemus, M., Valles, V., and de Alvarez-Buylla, E.R. (2003). Induction of brain glucose uptake by a factor secreted into cerebrospinal fluid. Brain Res. *994*, 124–133.

Barski, A., Cuddapah, S., Cui, K., Roh, T.Y., Schones, D.E., Wang, Z., Wei, G., Chepelev, I., and Zhao, K. (2007). High-resolution profiling of histone methylations in the human genome. Cell *129*, 823–837.

Birnbaum, K., Shasha, D.E., Wang, J.Y., Jung, J.W., Lambert, G.M., Galbraith, D.W., and Benfey, P.N. (2003). A gene expression map of the Arabidopsis root. Science *302*, 1956–1960.

Bolduc, N., Yilmaz, A., Mejia-Guerra, M.K., Morohashi, K., O'Connor, D., Grotewold, E., and Hake, S. (2012). Unraveling the KNOTTED1 regulatory network in maize meristems. Genes Dev. *26*, 1685–1690.

Bowman, J.L., Smyth, D.R., and Meyerowitz, E.M. (1989). Genes directing flower development in Arabidopsis. Plant Cell *1*, 37–52.

Brenner, S., Johnson, M., Bridgham, J., Golda, G., Lloyd, D.H., Johnson, D., Luo, S., McCurdy, S., Foy, M., Ewan, M., *et al.* (2000). Gene expression analysis by massively parallel signature sequencing (MPSS) on microbead arrays. Nat. Biotechnol. *18*, 630–634.

Busch, W., Miotk, A., Ariel, F.D., Zhao, Z., Forner, J., Daum, G., Suzaki, T., Schuster, C., Schultheiss, S.J., Leibfried, A., *et al.* (2010). Transcriptional control of a plant stem cell niche. Dev. Cell *18*, 849–861.

Charron, J.B., He, H., Elling, A.A., and Deng, X.W. (2009). Dynamic landscapes of four histone modifications during deetiolation in Arabidopsis. Plant Cell *21*, 3732–3748.

Coen, E.S., and Meyerowitz, E.M. (1991). The war of the whorls: genetic interactions controlling flower development. Nature *353*, 31–37.

Deal, R.B., and Henikoff, S. (2010). A simple method for gene expression and chromatin profiling of individual cell types within a tissue. Dev. Cell *18*, 1030–1040.

Deng, W., Ying, H., Helliwell, C.A., Taylor, J.M., Peacock, W.J., and Dennis, E.S. (2011). FLOWERING LOCUS C (FLC) regulates development pathways throughout the life cycle of Arabidopsis. Proc. Natl. Acad. Sci. U.S.A. *108*, 6680–6685.

Emmert-Buck, M.R., Bonner, R.F., Smith, P.D., Chuaqui, R.F., Zhuang, Z., Goldstein, S.R., Weiss, R.A., and Liotta, L.A. (1996). Laser capture microdissection. Science *274*, 998–1001.

Fodor, S.P., Rava, R.P., Huang, X.C., Pease, A.C., Holmes, C.P., and Adams, C.L. (1993). Multiplexed biochemical assays with biological chips. Nature *364*, 555–556.

Frommer, M., McDonald, L.E., Millar, D.S., Collis, C.M., Watt, F., Grigg, G.W., Molloy, P.L., and Paul, C.L. (1992). A genomic sequencing protocol that yields a positive display of 5-methylcytosine residues in individual DNA strands. Proc. Natl. Acad. Sci. U.S.A. *89*, 1827–1831.

Gehring, M., Bubb, K.L., and Henikoff, S. (2009). Extensive demethylation of repetitive elements during seed development underlies gene imprinting. Science *324*, 1447–1451.

Gilmour, D.S., and Lis, J.T. (1984). Detecting protein–DNA interactions *in vivo*: distribution of RNA polymerase on specific bacterial genes. Proc. Natl. Acad. Sci. U.S.A. *81*, 4275–4279.

He, G., Zhu, X., Elling, A.A., Chen, L., Wang, X., Guo, L., Liang, M., He, H., Zhang, H., Chen, F., *et al.* (2010). Global epigenetic and transcriptional trends among two rice subspecies and their reciprocal hybrids. Plant Cell *22*, 17–33.

He, G., Elling, A.A., and Deng, X.W. (2011). The epigenome and plant development. Annu. Rev. Plant Biol. *62*, 411–435.

Henderson, I.R., and Jacobsen, S.E. (2007). Epigenetic inheritance in plants. Nature *447*, 418–424.

Honma, T., and Goto, K. (2001). Complexes of MADS-box proteins are sufficient to convert leaves into floral organs. Nature *409*, 525–529.

Hsieh, T.F., Ibarra, C.A., Silva, P., Zemach, A., Eshed-Williams, L., Fischer, R.L., and Zilberman, D. (2009). Genome-wide demethylation of Arabidopsis endosperm. Science *324*, 1451–1454.

Huang, H., Mizukami, Y., Hu, Y., and Ma, H. (1993). Isolation and characterization of the binding sequences for the product of the Arabidopsis floral homeotic gene AGAMOUS. Nucleic Acids Res. *21*, 4769–4776.

Irish, V.F., and Sussex, I.M. (1990). Function of the apetala-1 gene during Arabidopsis floral development. Plant Cell *2*, 741–753.

Iyer, V.R., Horak, C.E., Scafe, C.S., Botstein, D., Snyder, M., and Brown, P.O. (2001). Genomic binding sites of the yeast cell-cycle transcription factors SBF and MBF. Nature *409*, 533–538.

Jiao, Y., and Deng, X.W. (2007). A genome-wide transcriptional activity survey of rice transposable element-related genes. Genome Biol. *8*, R28.

Jiao, Y., and Meyerowitz, E.M. (2010). Cell-type specific analysis of translating RNAs in developing flowers reveals new levels of control. Mol. Syst. Biol. *6*, 419.

Jofuku, K.D., den Boer, B.G., Van Montagu, M., and Okamuro, J.K. (1994). Control of Arabidopsis flower and seed development by the homeotic gene APETALA2. Plant Cell *6*, 1211–1225.

Kaufmann, K., Muino, J.M., Jauregui, R., Airoldi, C.A., Smaczniak, C., Krajewski, P., and Angenent, G.C. (2009). Target genes of the MADS transcription factor SEPALLATA3: integration of developmental and hormonal pathways in the *Arabidopsis* flower. PLoS Biol. *7*, e1000090.

Kaufmann, K., Pajoro, A., and Angenent, G.C. (2010a). Regulation of transcription in plants: mechanisms controlling developmental switches. Nat. Rev. Genet. *11*, 830–842.

Kaufmann, K., Wellmer, F., Muino, J.M., Ferrier, T., Wuest, S.E., Kumar, V., Serrano-Mislata, A., Madueno, F., Krajewski, P., Meyerowitz, E.M., *et al.* (2010b). Orchestration of floral initiation by APETALA1. Science *328*, 85–89.

Krizek, B.A., and Fletcher, J.C. (2005). Molecular mechanisms of flower development: an armchair guide. Nat. Rev. Genet. *6*, 688–698.

Kurimoto, K., Yabuta, Y., Ohinata, Y., Ono, Y., Uno, K.D., Yamada, R.G., Ueda, H.R., and Saitou, M. (2006). An improved single-cell cDNA amplification method for efficient high-density oligonucleotide microarray analysis. Nucleic Acids Res. *34*, e42.

La Rota, C., Chopard, J., Das, P., Paindavoine, S., Rozier, F., Farcot, E., Godin, C., Traas, J., and Moneger, F. (2011). A data-driven integrative model of sepal primordium polarity in Arabidopsis. Plant Cell *23*, 4318–4333.

Li, P., Ponnala, L., Gandotra, N., Wang, L., Si, Y., Tausta, S.L., Kebrom, T.H., Provart, N., Patel, R., Myers, C.R., *et al.* (2010). The developmental dynamics of the maize leaf transcriptome. Nat. Genet. *42*, 1060–1067.

Lister, R., O'Malley, R.C., Tonti-Filippini, J., Gregory, B.D., Berry, C.C., Millar, A.H., and Ecker, J.R. (2008). Highly integrated single-base resolution maps of the epigenome in Arabidopsis. Cell *133*, 523–536.

Longabaugh, W.J., Davidson, E.H., and Bolouri, H. (2005). Computational representation of developmental genetic regulatory networks. Dev. Biol. *283*, 1–16.

Lu, C., Tej, S.S., Luo, S., Haudenschild, C.D., Meyers, B.C., and Green, P.J. (2005). Elucidation of the small RNA component of the transcriptome. Science *309*, 1567–1569.

Mandel, M.A., Gustafson-Brown, C., Savidge, B., and Yanofsky, M.F. (1992). Molecular characterization of the Arabidopsis floral homeotic gene APETALA1. Nature *360*, 273–277.

Mortazavi, A., Williams, B.A., McCue, K., Schaeffer, L., and Wold, B. (2008). Mapping and quantifying mammalian transcriptomes by RNA-Seq. Nat. Methods *5*, 621–628.

van Mourik, S., van Dijk, A.D., de Gee, M., Immink, R.G., Kaufmann, K., Angenent, G.C., van Ham, R.C., and Molenaar, J. (2010). Continuous-time modeling of cell fate determination in Arabidopsis flowers. BMC Syst. Biol. *4*, 101.

Mustroph, A., Zanetti, M.E., Jang, C.J., Holtan, H.E., Repetti, P.P., Galbraith, D.W., Girke, T., and Bailey-Serres, J. (2009). Profiling translatomes of discrete cell populations resolves altered cellular priorities during hypoxia in Arabidopsis. Proc. Natl. Acad. Sci. U.S.A. *106*, 18843–18848.

Nagalakshmi, U., Wang, Z., Waern, K., Shou, C., Raha, D., Gerstein, M., and Snyder, M. (2008). The transcriptional landscape of the yeast genome defined by RNA sequencing. Science *320*, 1344–1349.

Nelson, T., Gandotra, N., and Tausta, S.L. (2008). Plant cell types: reporting and sampling with new technologies. Curr. Opin. Plant Biol. *11*, 567–573.

Ouyang, X., Li, J., Li, G., Li, B., Chen, B., Shen, H., Huang, X., Mo, X., Wan, X., Lin, R., *et al.* (2011). Genome-wide binding site analysis of FAR-RED ELONGATED HYPOCOTYL3 reveals its novel function in Arabidopsis development. Plant Cell *23*, 2514–2535.

Pelaz, S., Ditta, G.S., Baumann, E., Wisman, E., and Yanofsky, M.F. (2000). B and C floral organ identity functions require SEPALLATA MADS-box genes. Nature *405*, 200–203.

Ren, B., Robert, F., Wyrick, J.J., Aparicio, O., Jennings, E.G., Simon, I., Zeitlinger, J., Schreiber, J., Hannett, N., Kanin, E., *et al.* (2000). Genome-wide location and function of DNA binding proteins. Science *290*, 2306–2309.

Riechmann, J.L., Krizek, B.A., and Meyerowitz, E.M. (1996a). Dimerization specificity of *Arabidopsis* MADS domain homeotic proteins APETALA1, APETALA3, PISTILLATA, and AGAMOUS. Proc. Natl. Acad. Sci. U.S.A. *93*, 4793–4798.

Riechmann, J.L., Wang, M., and Meyerowitz, E.M. (1996b). DNA-binding properties of *Arabidopsis* MADS domain homeotic proteins APETALA1, APETALA3, PISTILLATA and AGAMOUS. Nucleic Acids Res. *24*, 3134–3141.

Robertson, G., Hirst, M., Bainbridge, M., Bilenky, M., Zhao, Y., Zeng, T., Euskirchen, G., Bernier, B., Varhol, R., Delaney, A., *et al.* (2007). Genome-wide profiles of STAT1 DNA association using chromatin immunoprecipitation and massively parallel sequencing. Nat. Methods *4*, 651–657.

Schena, M., Shalon, D., Davis, R.W., and Brown, P.O. (1995). Quantitative monitoring of gene expression patterns with a complementary DNA microarray. Science *270*, 467–470.

Schmid, M.W., Schmidt, A., Klostermeier, U.C., Barann, M., Rosenstiel, P., and Grossniklaus, U. (2012). A powerful method for transcriptional profiling of specific cell types in eukaryotes: laser-assisted microdissection and RNA sequencing. PLoS ONE *7*, e29685.

Shiraishi, H., Okada, K., and Shimura, Y. (1993). Nucleotide sequences recognized by the AGAMOUS MADS domain of Arabidopsis thaliana *in vitro*. Plant J. *4*, 385–398.

Slotkin, R.K., Vaughn, M., Borges, F., Tanurdzic, M., Becker, J.D., Feijo, J.A., and Martienssen, R.A. (2009). Epigenetic reprogramming and small RNA silencing of transposable elements in pollen. Cell *136*, 461–472.

Smaczniak, C., Immink, R.G., Muino, J.M., Blanvillain, R., Busscher, M., Busscher-Lange, J., Dinh, Q.D., Liu, S., Westphal, A.H., Boeren, S., *et al.* (2012). Characterization of MADS-domain transcription factor complexes in Arabidopsis flower development. Proc. Natl. Acad. Sci. U.S.A. *109*, 1560–1565.

Tang, F., Barbacioru, C., Wang, Y., Nordman, E., Lee, C., Xu, N., Wang, X., Bodeau, J., Tuch, B.B., Siddiqui, A., *et al.* (2009). mRNA-Seq whole-transcriptome analysis of a single cell. Nat. Methods *6*, 377–382.

Van Gelder, R.N., von Zastrow, M.E., Yool, A., Dement, W.C., Barchas, J.D., and Eberwine, J.H. (1990). Amplified RNA synthesized from limited quantities of heterogeneous cDNA. Proc. Natl. Acad. Sci. U.S.A. *87*, 1663–1667.

Wang, X., Elling, A.A., Li, X., Li, N., Peng, Z., He, G., Sun, H., Qi, Y., Liu, X.S., and Deng, X.W. (2009). Genome-wide and organ-specific landscapes of epigenetic modifications and their relationships to mRNA and small RNA transcriptomes in maize. Plant Cell *21*, 1053–1069.

Wang, Y., and Jiao, Y. (2011). Advances in plant cell type-specific genome-wide studies of gene expression. Front Biol. *6*, 384–389.

Wellmer, F., Alves-Ferreira, M., Dubois, A., Riechmann, J.L., and Meyerowitz, E.M. (2006). Genome-wide analysis of gene expression during early Arabidopsis flower development. PLoS Genet. *2*, e117.

Wellmer, F., and Riechmann, J.L. (2010). Gene networks controlling the initiation of flower development. Trends Genet. *26*, 519–527.

Wilhelm, B.T., Marguerat, S., Watt, S., Schubert, F., Wood, V., Goodhead, I., Penkett, C.J., Rogers, J., and Bahler, J. (2008). Dynamic repertoire of a eukaryotic transcriptome surveyed at single-nucleotide resolution. Nature *453*, 1239–1243.

Wuest, S.E., Vijverberg, K., Schmidt, A., Weiss, M., Gheyselinck, J., Lohr, M., Wellmer, F., Rahnenfuhrer, J., von Mering, C., and Grossniklaus, U. (2010). Arabidopsis female gametophyte gene expression map reveals similarities between plant and animal gametes. Curr. Biol. *20*, 506–512.

Yant, L., Mathieu, J., Dinh, T.T., Ott, F., Lanz, C., Wollmann, H., Chen, X., and Schmid, M. (2010). Orchestration of the floral transition and floral development in Arabidopsis by the bifunctional transcription factor APETALA2. Plant Cell *22*, 2156–2170.

Zanetti, M.E., Chang, I.F., Gong, F., Galbraith, D.W., and Bailey-Serres, J. (2005). Immunopurification of polyribosomal complexes of Arabidopsis for global analysis of gene expression. Plant Physiol. *138*, 624–635.

Zhang, C., Barthelson, R.A., Lambert, G.M., and Galbraith, D.W. (2008). Global characterization of cell-specific gene expression through fluorescence-activated sorting of nuclei. Plant Physiol. *147*, 30–40.

Zhou, J., Wang, X., He, K., Charron, J.B., Elling, A.A., and Deng, X.W. (2010). Genome-wide profiling of histone H3 lysine 9 acetylation and dimethylation in Arabidopsis reveals correlation between multiple histone marks and gene expression. Plant Mol. Biol. *72*, 585–595.

Next-generation Sequencing and the Future of Microbial Metagenomics

9

Andreas Wilke, Peter Larsen and Jack A. Gilbert

Abstract

Microbial ecology is going through a turbulent time, which is to be expected for a relatively young field. Here we explore the impact of one technique, microbial metagenomics, the study of sequencing DNA fragments from multiple microbial genomes in an assemblage of organisms, on the field of microbial ecology. We take a clear exploration of the reasons for looking at communities of microorganisms, and the potential for new discoveries of ecological importance, such as ubiquitous species distribution, and environmental community structuring factors. We explore the importance of community-driven research initiatives to help tackle the immensity of global microbial dynamics, and discuss the financial and practical implications of data generation and bioinformatics. Finally we highlight the role of microbial ecosystem modelling in helping to synthesize theories across domains. Microbial ecology has been, and will continue to be advanced through the application of careful experimental design and sophisticated technology. Metagenomics has a continuing role to play in this discovery, and as researchers ask more complex questions, this form of data generation will continue to help to answer those questions.

Introduction

Since bacteria were first identified there has been considerable interest regarding their diversity and ecology, most obviously in respect to human disease. Now they are recognized as vital to ecosystem function in any environment. Microbial communities, far from being a primitive form of life, are a sophisticated and highly networked super-organism. Although we know only a small amount about the *Bacteria,* they represent the oldest lineage on the tree of life, appearing on Earth approximated 3.8 billion years ago (Mojzsis *et al.*, 1996). Microbes are found in virtually every environment investigated to date, from boiling lakes (Barns *et al.*, 1994; Hugenholtz *et al.*, 1998) to the very rock itself (Fisk *et al.*, 2003). Given this ubiquity it is quite surprising to think that there are only a little over 15,000 recognized bacterial species, compared to nearly 200,000 described eukaryotic species (GOLD database – www.genomesonline.org). This paucity of available information is because there must be a cultured representative of each microbe to derive a biochemical description of its genetic and phenotypic characteristics (www.bergeys.org), and cultured representatives represent an estimated less than 2% of all bacteria (Hugenholtz and Pace, 1996).

With or without physical representatives, it is critical that we are able to define microbial diversity in any ecosystem. This task is made daunting primarily because the genotypic variation of microbial populations is vast. Considerable debate continues to rage around how to classify bacteria at the species level (Philippot *et al.*, 2010). Two principle methods for describing a microbial community are by its taxonomic diversity or its functional capacity. For *Bacteria* and *Archaea,* the predominant taxonomic determinant of a species has become the sequence of the 16S ribosomal RNA gene. However, it is clear that this gene is insufficient to accurately differentiate microbes at the species level for many of the recognized taxa.

In this chapter, we recognize the formal description of species as provided by Bergey's and accept the concept of relative levels of whole genome hybridization above 70% indicate speciation, whether between isolates or by virtual measures. However, it may be that classifying all bacteria at the species level is irrelevant to understanding the role of microbial diversity in an environment; rather it is more appropriate to consider the bacterial pan-genome, a species-cloud composed of groups of closely related bacteria (Medini *et al.*, 2005). Indeed through the recognized mechanisms and limitations to successful gene exchange (sex) it is likely that many bacteria maintain the cohesiveness of speciation (Sorensen *et al.*, 2005).

Tracking microbial diversity

It is important to ask, what benefit do we gain from a catalogue of microbial species diversity? Microbes are vital to humanity and the rest of the biosphere because of the services they provide, supporting every single aspect of biochemical interaction that maintains life on Earth. However, the whole concept of taxonomic diversity may be overall less important for microbes than measures of microbial functional diversity. It is therefore, potentially, more relevant to determine the functional capacity and regulation of that capacity in ecosystems at the level of the community, ignoring the monomers of individual cells or species. Admittedly this argument would not hold for long, as understanding cellular biochemistry and species interaction is absolutely necessary if we are to make sense of community interaction. Yet the surge of interest in microbial community structure and function, enabled in part by advances in sequencing technology, means that we must revisit an old discussion of what it is we are looking for in microbial communities. The old adage of 'Who is there?' and 'What do they do?' does not enable appropriate ecological network inference any more. We need to move past natural history and into community ecology, using defined interactions observed through in situ investigation to create working abstract models of community dynamics. This will be a cyclical endeavour, a constant feedback loop of improvement, where hypothesis defines model, defines hypothesis, refines model, defines hypothesis, ad infinitum.

The focus of this chapter is to discuss the use of next-generation sequencing technologies to explore and define the microbial world as a function of community. Here we will briefly define the past, present and explore the future of microbial metagenomics.

Metagenomics, is the study of multiple genomes (Gilbert and Dupont, 2011), defined as an exploration of total community DNA, including genomes and extra chromosomal elements from a random sampling of a community (Handelsman *et al.*, 1998). It is not currently financially or technically possible to sequence every genome in a defined community sample. The future of metagenomics seeks to remove the 'meta' from metagenomics. Genomics as applied to a community would require the complete genome sequence analysis of every singular cell, or representative of every microbial species in a system (Woyke *et al.*, 2009). Currently, cultured representatives of bacterial taxa have yielded more than 20,000 genomic sequences (www.genomes-online.org). Yet only 26 of the 61 known and candidate phyla have a representative cultured isolates with genome sequence (Hugenholtz and Kyrpides, 2009). Concerted efforts such as the Genomic Encyclopaedia of Bacteria and Archaea (GEBA – http://www.jgi.doe.gov/programs/GEBA/) and Microbial Earth (http://genome.jgi-psf.org/programs/bacteria-archaea/MEP/index.jsf) seek to rectify this disparity by greatly increasing not only the number of sequenced genomes, but also the taxonomic range of those genomes, to create a detailed phylogenomic picture of microbial life. It is accepted that the considered design of growth media will increase the number of microbes that can be cultured in the laboratory. Nonetheless the assessed diversity as understood from cultured strains remains an extreme underestimation, and new approaches for investigating the genetic diversity of microbial life are needed. One such approach, single-celled genomics, is still in its infancy. Single-cell genomics seeks to sequence an entire microbial genome from a single representative, non-cultured cell and this technique holds much promise for delivering on the ultimate goal of a fully characterized microbial ecosystem (Rodrigue *et al.*, 2009; Wang and Bodovitz, 2010; Woyke *et al.*, 2010; Zare and Kim, 2010; Grindberg *et al.*, 2011; Kalisky and Quake, 2011; Siegl *et al.*, 2011; Yoon *et al.*, 2011).

Applying omics technologies

The development of new 'omics based technologies to rapidly investigate the diversity of environmental microbial communities means it is now possible to start to characterize the relative abundance of all taxa and functions in a defined community (Jansson *et al.*, 2011). This will have profound consequences for microbial ecology. For example, it is possible that within 5 years, through shotgun metagenomics, single-celled community genomics, and 16S rRNA amplicons sequencing, we could finally test the hypothesis that 'everything is everywhere but the environment selects' (Whitfield, 2005; de Wit and Bouvier, 2006; O'Malley, 2007). The field of metagenomics has progressed from single-gene amplification, cloning and sequencing, to small and large genomic fragment cloning and sequencing, to shotgun sequencing of small fragments of community DNA in the space of 20 years. Recent advances in sequencing technology, outlined in chapters 1 and 2 of this volume, have only taken effect in the literature of environmental metagenomics since 2006 (Angly *et al.*, 2006; Edwards *et al.*, 2006; Krause *et al.*, 2006; Leininger *et al.*, 2006; Sogin *et al.*, 2006; Turnbaugh *et al.*, 2006), and all were related to the Roche 454 pyrosequencing platform. Prior to, and even following, this revolution, large-insert and small insert metagenomic library sequencing have dominated (reviewed by (Heidelberg *et al.*, 2010)). Indeed the most referenced metagenomic study to date is still the Global Ocean Sampling (Rusch *et al.*, 2007) survey which was small insert libraries sequenced using Sanger technology (for a comprehensive review of the discoveries from this survey read (Gilbert and Dupont, 2011).

The development of sequencing throughput has been astonishing since 2006, with new platforms (e.g. Illumina HiSeq2000) able to generate a trillion (1×10^{12}) base pairs of sequence data per run. This highlights interesting questions of coverage. To completely sequence all of the DNA in a community it will be necessary to either isolate the cellular and viral units in a system and sequence the chromosomal and extra-chromosomal material as described above, or to shotgun sequence all of the material to a depth of coverage adequate to accurately re-assemble the units from the set of data. Increasingly deeper sequencing efforts are now enabling genomic reassembly from metagenomic data (Hess *et al.*, 2011), just as viral genome and plasmid re-assembly have been accomplished at lower levels of metagenomic coverage (Williamson *et al.*, 2008; Palenik *et al.*, 2009; Jones *et al.*, 2010). However, a trillion base pairs is still a very small amount of DNA coverage relative to the amount and diversity of sequence in an environmental sample (Fig. 9.1). An average millilitre of seawater captured at the ocean's surface will contain 1×10^6 bacterial cells, based on a rule-of-thumb average of 2 million base pairs per genome gives an approximate quantity of 2 trillion base pairs of DNA. It could be argued that while Eukaryotic and archaeal biomass may not increase this significantly, viral genomic material probably would. The problem is multiplied in a gram of soil, which has approximately 1×10^9 cells with a higher average genome size (e.g. 3×10^6) producing approximately 3 quadrillion (3×10^{15}) base pairs of DNA. Hence, if you could isolate the DNA in a millilitre of water you could probably use 4 HiSeq2000 runs and get ~2× coverage of the whole metagenome. However, this is currently impossible, because the average marine bacterium has only a femtogram of genomic DNA (1×10^{-15} g), and hence a millilitre has only ~1 ng of DNA (1×10^{-9} g). This is not enough material to generate the sequencing libraries for any platform at this stage. Multiple displacement amplification to increase the yield of the DNA is one option, but there are significant concerns regarding the random bias associated with the amplification and what impact this would have upon the relative abundance of the genomes and genomic material in the original sample (Gilbert *et al.*, 2009). Obviously, based on the same assumption, a gram of soil could generate (in a perfect world) 3 micrograms of DNA, which is plenty for a metagenomic library on an Illumina HiSeq2000 library preparation. However, this would require 6000 HiSeq2000 runs. The current cost (February 2013) for an Illumina HiSeq2000 run with

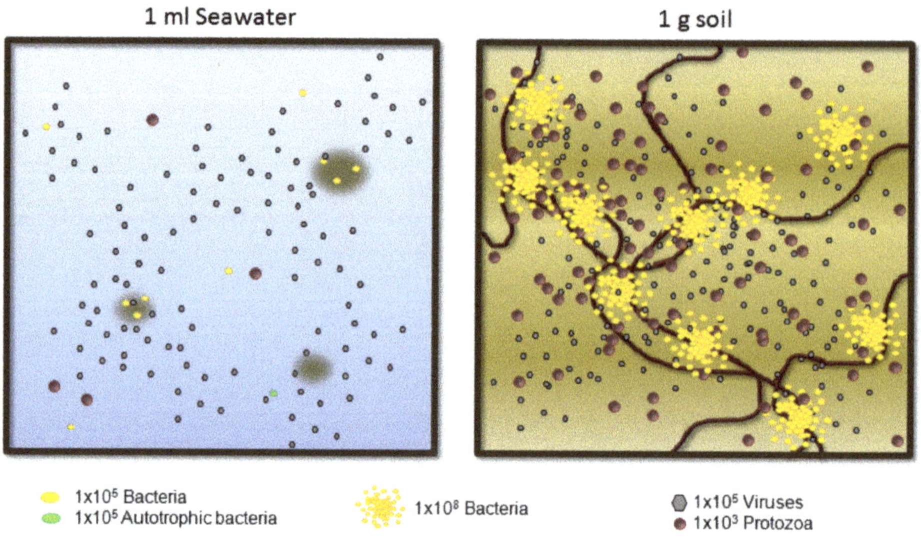

Figure 9.1 Representing the density and abundance of single cellular microorganisms in 1 ml of seawater and 1 g of soil. For the most part 1 ml of seawater contains disparately associated cells in a sea of potential viral predators. Occasional organic matter particles act as a temporary site for certain microbial species to attach and bloom in abundance. In 1 g of soil the bacterial community is potentially much more clustered around specific sites associated with appropriate nutrient requirements, leading to extensive compartmentalization of the community structure.

a 150 bp paired end run is \$44 per gigabase (1 billion base pairs) of sequence. Therefore, the theoretical cost of sequencing the DNA in a millilitre of seawater would be \$184,000, while for a gram of soil it would be approximately \$276 million. There are continuing improvements in the chemistry for library construction for each platform, Illumina HiSeq2000 library construction has been validated down to 50 ng of starting material using NEXTERA, although the heterogeneity of DNA quality means this is probably not yet of universal applicability. Observing past trends in sequencing capacity however provides some guide to our expectations for the future of metagenomic sequencing. In 1990, the Human Genome Project cost \$3 billion and was expected to take 15 years to complete. Twenty years later, the National Human Genome Research Institute currently lists the cost of sequencing a human genome at \$10,000. Assuming a similar, linear trend for performance of sequencing technologies over the next 20 years, all of the DNA in a gram of soil will cost less than \$1000. The advances of sequencing technology have not been linear however; they have been exponential. It is not unreasonable to expect that it will not be long until complete DNA sequencing of environmental samples is not only an achievable, but relatively trivial task.

The question should be asked though, 'do we really need to sequence everything in a community?' The ultimate answer is 'yes'. The Earth's biogeochemical cycles relies on the interactions between the microbial populations. Our ability to maintain our place in this ecosystem depends on our understanding of these interrelationships. Investigation of the relationships is complicated as microbial diversity is not necessarily equivalent to microbial metabolic capacity. Therefore, a shallow analysis of the metagenomic potential of the system should provide enough coverage of these abundant groups to enable appropriate ecological inference. In fact this is the case with every single metagenomic study to date. However, recent evidence from the Western English Channel long-term monitoring station L4 has shown that this ecosystem had a persistent microbial community (Caporaso *et al.*, 2012; Gibbons *et al.*, 2013). This was achieved by sequencing 10,000 reads using 16S rRNA amplicon pyrosequencing from 72 consecutive time points over 6 years in the surface water (Gilbert *et al.*, 2012), then one amplicon sample was randomly picked (December 2007) and sequenced to a depth of 10 million 16S rDNA reads. This demonstrated that 99.96% of the taxa identified in the 72 10,000 read samples were found in the one deep sequenced sample. More explicitly, the total diversity found in the 72 time points comprised only 5% of the taxonomic richness found in the deep sequenced dataset. This calls into question the determination of what constitutes an abundant microbe. In a previous manuscript (Gilbert *et al.*, 2009) rare bacteria were defined as those whose relative average abundance comprised <0.1% of total population. We would therefore potentially dismiss these microbes as not being of significant importance in the community. However, when we sequence to a depth of 10 million reads, these once 'rare' bacteria, are suddenly relatively far more abundant. At this level of sequencing coverage, they comprise the top 5% of the community at any given time in the annual cycle. Hence, 'rare' is relative to depth of sequencing. To have a system whereby you could absolutely quantify the abundance of every taxon would provide an opportunity to track the relative importance of every organism, especially as a number of 'rare' taxa have been observed to bloom at irregular intervals to become a large if not majority member of the microbial community diversity.

Depth of coverage and the deliberation between rare and abundant microbes is very important to understanding ecosystem dynamics and community structure. However, it is potentially less important for functional dynamics. Fundamentally, most microorganisms have a suite of core 'housekeeping' functions. This becomes very obvious when we compare different 'omic technologies for the same ecosystem. The same community can show dramatic variations in the relative abundance of the higher-level taxa when explored using 16S rRNA amplicon sequencing, e.g. through time (Gilbert *et al.*, 2012) or space (Herlemann *et al.*, 2011), and yet show very little change in the higher level functional capacity (Gilbert *et al.*, 2010). When exploring the transcriptional response of that capacity, e.g. using metatranscriptomics, it is again possible to see considerable changes through time as the level of gene expression alters to suite the environmental conditions (Gilbert *et al.*, 2010). Therefore, it could be argued that deep surveys of functional capacity may not adequately describe the differences in ecological interaction through time, as these could be masked by the mass of 'house-keeping' functions. Therefore, metatranscriptomic analysis should provide a better window into real differences in the system, especially as 'housekeeping' transcripts are generally lower in abundance than transcripts for genes relating to changing physiological state, e.g. nutritional status (Gilbert *et al.*, 2008). Real time measurements of metatranscriptomes from in situ collected samples are becoming a reality, so that transcriptional profiles can be obtained at multiple time scales ((Feike *et al.*, 2011; Ottesen *et al.*, 2011). Importantly, as metatranscriptomes generally require a lower depth of sequencing to acquire greater coverage of function in a community (Gilbert, 2010), this approach can utilize current sequencing depth to achieve excellent coverage of the system, although complete coverage will still require considerable sequencing investments (Gifford *et al.*, 2011).

Designing experiments

The start of any study is a good hypothesis (Gilbert *et al.*, 2011). Historically, cost limitations associated with metagenomic sequencing have had an impact on the quality of the experimental design; essentially available budget for most independent researchers did not allow for traditional statistically relevant experimental design. One of the first sacrifices to be made in the race to use metagenomic-sequencing technologies was replication. Without biological replicates with which to perform non-parametric statistical assessment of within sample variance, many researchers fell back on bootstrapping techniques to provide some measure of significance. Only recently has the effective cost of sequencing dropped to a level whereby researchers can include replication. This will become increasingly prevalent as sequencing costs drop, and the funding paradigm shifts with increasing realization of the importance of microbial community characterization to all aspects of planetary science. As budget awards start to match expectations it will become easier and easier to not compromise the scientific principle and statistically relevant experimental design. The next likely limitation on designing and analysing metagenomic sequencing experiments will be computing power. The cost of sequencing is dropping far more rapidly than the cost of computational capacity. Rather than limit future experiments by ability to acquire sequence data, they will be limited in the ability to store, analyse, and interpret the data collected.

As metagenomics moves forward as field, it is growing in maturity; internationally cooperative initiatives such as the Earth Microbiome Project (www.earthmicrobiome.org) are providing collaborative frameworks in which to enable the generation of comparable cohesive datasets of tens of thousands of environmental samples (Gilbert *et al.*, 2010a,b, 2011). The Genomic Standards Consortium (www.gensc.org) is creating appropriate checklists to ensure researchers know which data about their sampled environment and what information about the sample processing, nucleic acid extraction, amplification, sequencing, and bioinformatics analysis should be recorded, and importantly, how they should be recorded to ensure that the data are usable by the extended community. (Field *et al.*, 2011; Yilmaz *et al.*, 2011a,b). Additionally, online metagenomic analysis portal such as MG-RAST (Meyer *et al.*, 2008) are enabling democratization of bioinformatic analysis capability. This resource (http://metagenomics.anl.gov) is free, and currently hosts more than 7000 users, from 72 countries, with data generated by projects funded by more than 100 agencies around the world. To date it has processed over 3 trillion base pairs of metagenomic data, and routinely allows more than 500 users to process upwards of 40 billion base pairs of metagenomic information a day. While the call to have a bioinformatician in every university department may still be valid, the computing resources required to handle big datasets may be more easily processed using the experience and capability already available online.

Modelling microbial diversity

With the perceived grunt work of metagenomic data screening, annotation and analysis provided externally, researchers are free to explore the unique capability of their own datasets; and if generated in comparable frameworks, such as the EMP, then the extended capabilities of their data and every other compliant dataset as well. One of the key developments for both understanding and manipulating microbial communities in complex ecosystems is the use of abstract mathematical models, whereby community dynamics are tied to environmental and climatic characteristics. The development of bioclimatic models to enable prediction of species' ranges as a function of environment and climate parameters has been driven by a need to understand changing ecosystem services in a changing world. It is essential that humanity can predict how the changes in the natural world will affect them, so than mitigation of negative outcomes can prevent economic and socio-political disasters. There are a number of bioclimatic model methods currently in use that utilize environmental data (Heikkinen, 2006; Jeschke and Strayer, 2008; Follows and Dutkiewicz, 2011), defining an organism's habitable range as a function of environmental parameters. However, while each has strengths, none are currently capable of appropriately modelling the extraordinary dynamics of microbial communities. Logistic regression uses generalized linear models (GLM) (Bolker, 2009) to regress the presence or absence of a species against climatic variables as a linear function. Generalized additive models (GAM) model species as an additive combination of functions for independent variables (Hastie and Tibshirani, 1990). Climate envelope models [e.g. BIOCLIM (Busby, 1991), DOMAIN (Carpenter, 1993), HABITAT (Walker, 1991)] fit the minimal envelope that defines an organism's possible habitat in multi-dimensional space, but use presence only data rather than presence/absence. The classification and regression tree (CART) technique, models populations as a binary decision tree in which the decision rules at each node use one or more independent environmental parameter variables. Neural Network approaches such as genetic algorithm for rule-set prediction (GARP) (Stockwell, 1992, 1999) have powerful predictive capabilities, but only models organism distributions as present or absent as a function of environmental parameters. These bioclimatic models mostly disregard complex biotic interactions. Recently, a new bioclimatic model microbial assemblage prediction (MAP; Larsen *et al.*, 2011a) was developed off the back of extensive metagenomic characterization of a coastal marine ecosystem, which explicitly models interactions between microbial taxa. Importantly, MAP also enables extrapolative prediction over time and space of taxon abundance in a microbial community. Linking this novel technology to existing metabolic network predictive tools [e.g. predictive relative metabolic turnover (PRMT) Larsen *et al.* (2011b)] provides the opportunity to use remote and in situ environmental parameter measurements to predict not only the taxonomic community structure, but also the metabolic capability of the community. PRMT uses metagenomic or metatranscriptomic data and predicts a metabolic interaction network that enables researchers to predict where and when a specific metabolite (e.g. CO_2) will be relatively more consumed or produced in one microbial population when compared to another. By linking this to predictive modelling infrastructure such as MAP it is possible to manipulate environmental conditions *in silico* to drive predictions of whether specific areas of terrestrial or aquatic ecosystems will become carbon sources or carbon sinks in reaction to changing weather patterns or climate.

Concluding remarks

The future of metagenomics lies in more comprehensive characterization of microbial community structure, improvements in characterization of taxonomic and functional composition in metagenomics data, reduced budget constraints enabling statistically relevant experimental design to become manifest, increased reliance of online data analysis portals, and the introduction of microbial dynamics into local, regional, continental and global scale climate models. We stand on a precipice of capability, and only as a cohesive cooperative can we handle the colossal tasks ahead. However, the future is bright.

Acknowledgement

This work was supported by the US Department of Energy under Contract DE-AC02-06CH11357.

References

Angly, F.E., Felts, B., Breitbart, M., Salamon, P., Edwards, R.A., Carlson, C., Chan, A.M., Haynes, M., Kelley, S., Liu, H., *et al.* (2006). The marine viromes of four oceanic regions. PLoS Biol. *4*(11), e368.

Barns, S.M., Fundyga, R.E., Jeffries, M.W., and Pace, N.R. (1994). Remarkable archaeal diversity detected in a Yellowstone National Park hot spring environment. Proc. Natl. Acad. Sci. U.S.A. *91*(5), 1609–1613.

Bolker, B.M., Clark, C.J., Geange, S.W., Poulsen, J.R., Stevens, M.H., and White, J.S. (2009). Generalized linear mixed models: a practical guide for ecology and evolution. Trends Ecol. Evol. 3(24), 171–193.

Busby, J.R. (1991). BIOCLIM – A Bioclimatic Analysis and Prediction System. In Nature Conservation: Cost Effective Biological Surveys and Data Analysis, Margules, R., and Austin, M.P., eds. CSIRO, Australia, pp. 64–68.

Caporaso, J.G., Paszkiewicz, K., Field, D., Knight, R., and Gilbert, J.A. (2012). Evidence for a persistent microbial community in the Western English Channel. ISME J. *6*, 1089–1093.

Carpenter, G., Gillison, A.N., and Winter, J. (1993). DOMAIN: a flexible modelling procedure for mapping potential distributions of plants and animals. Biodiversity and Conservation 2, 667–680.

Edwards, R.A., Rodriguez-Brito, B., Wegley, L., Haynes, M., Breitbart, M., Peterson, D.M., Saar, M.O., Alexander, S., Alexander, E.C., and Rohwer, F. (2006). Using pyrosequencing to shed light on deep mine microbial ecology. BMC Genomics 7.

Feike, J., Jurgens, K., Hollibaugh, J.T., Kruger, S., Jost, G., and Labrenz, M. (2011). Measuring unbiased metatranscriptomics in suboxic waters of the central Baltic Sea using a new in situ fixation system. ISME J.

Field, D., Amaral-Zettler, L., Cochrane, G., Cole, J.R., Dawyndt, P., Garrity, G.M., Gilbert, J., Glockner, F.O., Hirschman, L., Karsch-Mizrachi, I., *et al.* (2011). The Genomic Standards Consortium. Plos Biology 9(6).

Fisk, M.R., Storrie-Lombardi, M.C., Douglas, S., Popa, R., McDonald, G., and Di Meo-Savoie, C. (2003). Evidence of biological activity in Hawaiian subsurface basalts. Geochemistry Geophysics Geosystems *4*.

Follows, M.J., and Dutkiewicz, S. (2011). Modeling diverse communities of marine microbes. Annual Review Marine Science 3, 427–451.

Gibbons, S.M., Caporaso, J.G., Pirrung, M., Field, D., Knight, R., and Gilbert, J.A. (2013). Evidence for a persistent microbial seed bank throughout the global ocean. PNAS (In press).

Gifford, S.M., Sharma, S., Rinta-Kanto, J.M., and Moran, M.A. (2011). Quantitative analysis of a deeply sequenced marine microbial metatranscriptome. ISME J. *5*(3), 461–472.

Gilbert, J.A. (2010). Beyond the Infinite – tracking bacterial gene expression. Microbiol. Today 37(2), 82–85.

Gilbert, J.A., and Dupont, C.L. (2011). Microbial metagenomics: beyond the genome. Annu. Rev. Mar. Sci. 3, 347–371.

Gilbert, J.A., Field, D., Huang, Y., Edwards, R., Li, W., Gilna, P., and Joint, I. (2008). Detection of large numbers of novel sequences in the metatranscriptomes of complex marine microbial communities. PLoS One *3*(8), e3042.

Gilbert, J.A., Field, D., Swift, P., Newbold, L., Oliver, A., Smyth, T., Somerfield, P.J., Huse, S., and Joint, I. (2009a). The seasonal structure of microbial communities in the Western English Channel. Environ. Microbiol. *11*(12), 3132–3139.

Gilbert, J.A., Zhang, K., and Neufeld, J.D. (2009b). Multiple Displacement Amplification. In Handbook of Hydrocarbon and Lipid Microbiology, Timmis, K. ed. (Berlin, Springer-Verlag). *1*, pp. 4255–4263.

Gilbert, J.A., Field, D., Swift, P., Thomas, S., Cummings, D., Temperton, B., Weynberg, K., Huse, S., Hughes, M., Joint, I., *et al.* (2010a). The taxonomic and functional diversity of microbes at a temperate coastal site: a 'multi-omic' study of seasonal and diel temporal variation. PLoS One *5*(11), e15545.

Gilbert, J.A., Meyer, F., Antonopoulos, D., Balaji, P., Brown, C.T., Desai, N., Eisen, J.A., Evers, D., Field, D., Feng, W., *et al.* (2010b). Meeting report: the terabase metagenomics workshop and the vision of an Earth microbiome project. Stand. Genomic Sci. 3(3), 243–248.

Gilbert, J.A., Meyer, F., Jansson, J., Gordon, J., Pace, N., Tiedje, J., Ley, R., Fierer, N., Field, D., Kyrpides, N., *et al.* (2010c). The Earth Microbiome Project: Meeting report of the '1 EMP meeting on sample selection and acquisition' at Argonne National Laboratory October 6 2010. Stand Genomic Sci 3, 249–253.

Gilbert, J.A., Meyer, F., and Bailey, M.J. (2011a). The future of microbial metagenomics (or is ignorance bliss?). ISME J. *5*, 777–779.

Gilbert, J.A., O'Dor, R.K., and Vogel, T.M. (2011b). Survey studies are still vital to science. Nature *469*, 162.

Gilbert, J.A., Steele, J.A., Caporaso, J.G., Steinbruck, L., Reeder, J., Temperton, B., Huse, S., McHardy, A.C., Knight, R., Joint, I., *et al.* (2012). Defining seasonal marine microbial community dynamics. ISME J. *6*, 298–308.

Grindberg, R.V., Ishoey, T., Brinza, D., Esquenazi, E., Coates, R.C., Liu, W.T., Gerwick, L., Dorrestein, P.C., Pevzner, P., Lasken, R., *et al.* (2011). Single cell genome amplification accelerates identification of the apratoxin biosynthetic pathway from a complex microbial assemblage. PLoS One *6*(4), e18565.

Handelsman, J., Rondon, M.R., Brady, S.F., Clardy, J., and Goodman, R.M. (1998). Molecular biological access to the chemistry of unknown soil microbes: a new frontier for natural products. Chem. Biol. *5*, R245–249.

Hastie, T., and Tibshirani, R. (1990). Exploring the nature of covariate effects in the proportional hazards model. Biometrics *46*, 1005–1016.

Heidelberg, K.B., Gilbert, J.A., and Joint, I. (2010). Marine genomics: at the interface of marine microbial ecology and biodiscovery. Microbiol. Biotechnology 3(5), 531–543.

Herlemann, D.P., Labrenz, M., Jurgens, K., Bertilsson, S., Waniek, J.J., and Andersson, A.F. (2011). Transitions in bacterial communities along the 2000 km salinity gradient of the Baltic Sea. ISME J.

Hess, M., Sczyrba, A., Egan, R., Kim, T.W., Chokhawala, H., Schroth, G., Luo, S., Clark, D.S., Chen, F., Zhang, T., *et al.* (2011). Metagenomic discovery of biomass-degrading genes and genomes from cow rumen. Science *331*(6016), 463–467.

Hugenholtz, P., and Kyrpides, N.C., (2009). A changing of the guard. Environ. Microbiol. *11*(3), 551–553.

Hugenholtz, P., and Pace, N.R. (1996). Identifying microbial diversity in the natural environment: a molecular phylogenetic approach. Trends_Biotechnol. *14*(6), 190–197.

Hugenholtz, P., Pitulle, C., Hershberger, K.L., and Pace, N.R. (1998). Novel division level bacterial diversity in a Yellowstone hot spring. J. Bacteriol. *180*(2), 366–376.

Jansson, J.K., Neufeld, J.D., Moran, M.A., and Gilbert, J.A. (2011). Omics for understanding microbial functional dynamics. Environ. Microbiol.

Jeschke, J.M., and Strayer, D.L. (2008). Usefulness of bioclimatic models for studying climate change and invasive species. Ann. New York Acad. Sci. *1134*, 1–24.

Jones, B.V., Sun, F., and Marchesi, J.R. (2010). Comparative metagenomic analysis of plasmid encoded functions in the human gut microbiome. BMC Genomics *11*, 46.

Kalisky, T., and Quake, S.R. (2011). Single-cell genomics. Nature Methods *8*, 311–314.

Krause, L., Diaz, N.N., Bartels, D., Edwards, R.A., Puhler, A., Rohwer, F., Meyer, F., and Stoye, J. (2006). Finding novel genes in bacterial communities isolated from the environment. Bioinformatics *22*(14), e281–289.

Larsen, P.E., Field, D., and Gilbert, J.A. (2011a). Predicting bacterial community assemblages using an artificial neural network approach. Nat. Methods *9*, 621–625.

Larsen, P., Collart, F., Field, D., Meyer, F., Keegan, K., C.S., H., McGrath, J.W., Quinn, J.P., and Gilbert, J.A. (2011b). Predicted Relative Metabolomic Turnover (PRMT): determining metabolic turnover from a coastal marine metagenomic dataset. BMC Microbial. Informatics and Experimentation *1*,4.

Leininger, S., Urich, T., Schloter, M., Schwark, L., Qi, J., Nicol, G.W., Prosser, J.I., Schuster, S.C., and Schleper, C. (2006). Archaea predominate among ammonia-oxidizing prokaryotes in soils. Nature *442*, 806–809.

Medini, D., Donati, C., Tettelin, H., Masignani, V., and Rappuoli, R. (2005). The microbial pan-genome. Curr. Opin. Genet. Dev. *15*(6), 589–594.

Meyer, F., Paarmann, D., D'Souza, M., Olson, R., Glass, E.M., Kubal, M., Paczian, T., Rodriguez, A., Stevens, R., Wilke, A., *et al.* (2008). The metagenomics RAST server – a public resource for the automatic phylogenetic and functional analysis of metagenomes. BMC Bioinformatics *9*, 386.

Mojzsis, S.J., Arrhenius, G., McKeegan, K.D., Harrison, T.M., Nutman, A.P., and Friend, C.R. (1996). Evidence for life on Earth before 3,800 million years ago. Nature *384*(6604), 55–59.

O'malley, M.A. (2007). The nineteenth century roots of 'everything is everywhere'. Nature Rev. Microbiol. *5*(8), 647–651.

Ottesen, E.A., Marin, R., 3rd, Preston, C.M., Young, C.R., Ryan, J.P., Scholin, C.A., and Delong, E.F. (2011). Metatranscriptomic analysis of autonomously collected and preserved marine bacterioplankton. ISME J.

Palenik, B., Ren, Q., Tai, V., and Paulsen, I.T. (2009). Coastal Synechococcus metagenome reveals major roles for horizontal gene transfer and plasmids in population diversity. Environ. Microbiol. *11*(2), 349–359.

Philippot, L., Andersson, S.G.E., Battin, T.J., Prosser, J.I., Schimel, J.P., Whitman, W.B., and Hallin, S. (2010). The ecological coherence of high bacterial taxonomic ranks. Nature Rev. Microbiol. *8*(7), 523–529.

Risto, K., Heikkinen, M.L., Miguel, B., Virkkala, R.A., Thuiller, W., and Sykes, M.T. (2006). Methods and uncertainties in bioclimatic envelope modeling under climate change. Prog. Phys. Geogr. *30*, 751–777.

Rodrigue, S., Malmstrom, R.R., Berlin, A.M., Birren, B.W., Henn, M.R., and Chisholm, S.W. (2009). Whole genome amplification and de novo assembly of single bacterial cells. PLoS One *4*, e6864.

Rusch, D.B., Halpern, A.L., Sutton, G., Heidelberg, K.B., Williamson, S., Yooseph, S., Wu, D., Eisen, J.A., Hoffman, J.M., Remington, K., *et al.* (2007). The Sorcerer II Global Ocean Sampling expedition: northwest Atlantic through eastern tropical Pacific. PLoS Biol. *5*(3), e77.

Siegl, A., Kamke, J., Hochmuth, T., Piel, J., Richter, M., Liang, C., Dandekar, T., and Hentschel, U. (2011). Single-cell genomics reveals the lifestyle of Poribacteria, a candidate phylum symbiotically associated with marine sponges. ISME J. *5*, 61–70.

Sogin, M.L., Morrison, H.G., Huber, J.A., Mark Welch, D., Huse, S.M., Neal, P.R., Arrieta, J.M., and Herndl, G.J. (2006). Microbial diversity in the deep sea and the underexplored 'rare biosphere'. Proc. Natl. Acad. Sci. U.S.A. *103*, 12115–12120.

Sorensen, S.J., Bailey, M., Hansen, L.H., Kroer, N., and Wuertz, S. (2005). Studying plasmid horizontal transfer in situ: a critical review. Nature Rev. Microbiol. 3(9), 700–710.

Stockwell, D.R.B., and Noble, I.R. (1992). Induction of sets of rules from animal distribution data: A robust and informative method of analysis. Math. Comp. Simulation 33, 385–390.

Stockwell, D.R.B., and Peters, D.P. (1999). The GARP modelling system: Problems and solutions to automated spatial prediction. Int. J. Geog. Inform. Syst. *13*, 143–158.

Turnbaugh, P.J., Ley, R.E., Mahowald, M.A., Magrini, V., Mardis, E.R., and Gordon, J.I. (2006). An obesity-associated gut microbiome with increased capacity for energy harvest. Nature *444*(7122): 1027–1031.

Walker, P.A., and Cocks, K.D. (1991). HABITAT: a procedure for modelling a disjoint environmental envelope for a plant or animal species. Global Ecol. Biogeog. Letters *1*, 108–118.

Wang, D.J., and Bodovitz, S. (2010). Single cell analysis: the new frontier in 'omics'. Trends Biotechnol. *28*, 281–290.

Whitfield, J. (2005). Biogeography. Is everything everywhere? Science *310*, 960–961.

Williamson, S.J., Rusch, D.B., Yooseph, S., Halpern, A.L., Heidelberg, K.B., Glass, J.I., Andrews-Pfannkoch, C., Fadrosh, D., Miller, C.S., Sutton, G., *et al.* (2008). The Sorcerer II Global Ocean Sampling Expedition: metagenomic characterization of viruses within aquatic microbial samples. PLoS One 3(1), e1456.

de Wit, R., and Bouvier, T.,(2006). 'Everything is everywhere, but, the environment selects'; what did Baas Becking and Beijerinck really say? Environ. Microbiol. *8*(4), 755–758.

Woyke, T., Xie, G., Copeland, A., Gonzalez, J.M., Han, C., Kiss, H., Saw, J.H., Senin, P., Yang, C., Chatterji, S., *et al.* (2009). Assembling the marine metagenome, one cell at a time. PLoS One *4*(4), e5299.

Woyke, T., Tighe, D., Mavromatis, K., Clum, A., Copeland, A., Schackwitz, W., Lapidus, A., Wu, D., McCutcheon, J.P., McDonald, B.R., *et al.* (2010). One bacterial cell, one complete genome. PLoS One 5(4), e10314.

Yilmaz, P., Gilbert, J.A., Knight, R., Amaral-Zettler, L., Karsch-Mizrachi, I., Cochrane, G., Nakamura, Y., Sansone, S.A., Glockner, F.O., and Field, D. (2011a). The genomic standards consortium: bringing standards to life for microbial ecology. ISME J. *5*, 1565–1567.

Yilmaz, P., Kottmann, R., Field, D., Knight, R., Cole, J.R., Amaral-Zettler, L., Gilbert, J.A., Karsch-Mizrachi, I., Johnston, A., Cochrane, G., *et al.* (2011b). Minimum information about a marker gene sequence (MIMARKS) and minimum information about any (x) sequence (MIxS) specifications. Nat. Biotech. *29*(5), 415–420.

Yoon, H.S., Price, D.C., Stepanauskas, R., Rajah, V.D., Sieracki, M.E., Wilson, W.H., Yang, E.C., Duffy, S., and Bhattacharya, D. (2011). Single-cell genomics reveals organismal interactions in uncultivated marine protists. Science 332(6030), 714–717.

Zare, R.N., and Kim, S. (2010). Microfluidic platforms for single-cell analysis. Annu. Rev. Biomed. Eng. *12*, 187–201.

Next-generation Sequencing, Metagenomes and the Human Microbiome

10

Karen E. Nelson, Ramana Madupu, Sebastian Szpakowski, Johannes B. Goll, Konstantinos Krampis and Barbara A. Methé

Abstract

Since 2003 there has been an explosion in the quantity of sequencing data generated through the study of whole microbial communities and their genetic content. These metagenomic surveys are accelerating the field of *Microbial Ecology*, revealing new species, pathways, functional capabilities, lateral gene transfer and enabling studies of microbial evolution. Venter and colleagues embarked on the first large-scale metagenomic study with the surveys of the Sargasso Sea, and the subsequent years have delivered significant improvements in high throughput sequencing technologies, coupled with a reduction of sequencing costs, which collectively has accelerated the rate and ease with which a range of environments are being studied with this approach. The end result has been a democratization of sequencing and now both large and small labs as well as individual investigators are capable of high-throughput metagenomic studies on any environment of choice. However, management, analysis and interpretation of these datasets remain an on-going challenge requiring continual improvement as technologies progress.

Introduction

Metagenomics employs high-throughput sequencing approaches to study a microbial environment or community. Metagenomics and associated 'omics' approaches have become routine as new sequencing technologies have increased data output and resulted in a reduction of the associated costs of generating these large datasets. The result has been a true revolution in how we approach biology, microbiology, microbial ecology and the study of uncultivated microbes. To date, tens of studies have been performed with samples collected from diverse habitats including the oceans (Venter *et al.*, 2004; Rusch *et al.*, 2007; Yooseph *et al.*, 2007), soil (Berlemont *et al.*, 2011; Reddy *et al.*, 2012), air (Tringe *et al.*, 2008; Whon *et al.*, 2012), human (Gill *et al.*, 2006), and animal hosts (Yildirim *et al.*, 2010). Some of these investigations have become national efforts (for example the National Institutes of Health funded Human Microbiome Project (HMP; http://commonfund.nih.gov/hmp/; NIH HMP Working Group *et al.*, 2009; Human Microbiome Jumpstart Reference Strains Consortium *et al.*, 2010; Human Microbiome Consortium, 2012a,b) and the Global Ocean Sampling Expedition; as well as international projects (MetaHIT; http://www.metahit.eu/; Qin *et al.*, 2010; Arumugam *et al.*, 2011), aimed at elucidating the biogeography of microbial communities and the ecology of the respective environments that are being investigated. The analysis of the human microbiome for example has brought about a paradigm shift in our understanding of the role of resident microorganisms in human health and disease, and brings non-traditional research approaches to the forefront of personalized medicine.

The true capabilities of new sequencing technologies and progress in the field was exemplified by the first large-scale environmental metagenomic study, conducted by J. Craig Venter and his colleagues at the J. Craig Venter Institute (JCVI), and chronicled in 2004 (Venter *et al.*, 2004) where the team reported the unprecedented discovery of more than 1.2 million new genes. This initial study evolved to become JCVI's Global Ocean Sampling (Venter *et al.*, 2004) initiative, which to date has generated millions of sequences derived from sites in the oceans worldwide. The Global Ocean Sampling expedition remains the largest protein dataset (Venter *et al.*, 2004; Rusch *et al.*, 2007; Yooseph *et al.*, 2007), and has also resulted in a number of spin-off studies that have been focused on the datasets that were initially generated (Williamson *et al.*, 2008; Wu *et al.*, 2011; Dupont *et al.*, 2012).

The animal species that have been subjected to metagenomic sequencing usually are investigated with a goal towards understanding the microbial populations associated with animal health and how they change under disease conditions and/or for identifying novel metabolisms. Investigations of ruminal microbial communities were among the earliest metagenomic studies. Microbial changes in the rumen under different dietary regimens revealed far more functional and species diversity in the rumen than previously appreciated (Brulc *et al.*, 2009). Studies with beef cattle suggest large variations in microbial populations that cannot be easily attributed to differences in age, diet or animal breed. However, these studies do suggest that there is a core set of microbial species shared across both beef and dairy cattle animals that includes those in phylum *Bacteroidetes* (genera *Prevotella* and *Bacteroides*); phylum *Firmicutes* (genera *Faecalibacterium, Ruminococcus, Roseburia,* and *Clostridium*); and phylum Proteobacterium genus *Succinovibrio* (Durso *et al.*, 2010). However, the relative abundances of these species are thought to differ between the two groups of cattle (Durso *et al.*, 2010). Several herbivorous species have also been examined for novel enzymatic activities associated with the metabolism of cellulose (and other plant) material. Pope and colleagues (Pope *et al.*, 2012) for example interrogated Svalbard reindeer where metagenomes were surveyed to reveal several polysaccharide utilization loci-like systems (PULs) as well as glycoside hydrolase and other carbohydrate-degrading enzyme families. Studies on the gastrointestinal tracts of non-human primates [black-and-white colobus (*Colubus guereza*), red colobus (*Piliocolobus tephrosceles*) and red-tailed guenon (*Cercopithecus ascanius*)] revealed that microbial community compositions among individuals within the same primate species are more similar to each other than to those of different primate species, showing that gastrointestinal microbiomes of wild primates co-vary with their hosts, likely a result of host genetics, the inherited microbiome and dietary influences (Yildirim *et al.*, 2010).

Marker-specific microbial community surveys

Diversity profiling using 16S rRNA gene sequences as a taxonomic marker has been an attractive approach for several reasons. This gene is present in all bacteria and archaea as a member of a core set of conserved informational genes that are believed to be transmitted most frequently by vertical descent (as opposed to horizontal transfer) (Daubin *et al.*, 2003). While parts of the 16S rRNA gene remained evolutionarily conserved under strong functional constraints, there also exist multiple hyper-variable regions subject to mutations and evolutionary drifts. From a technical standpoint we can readily use this gene sequence to interrogate microbial diversity across a large portion of inhabitants of an environment. Sequence similarity clustering [see Sun *et al.* (2012) for a benchmark and review of various approaches] using a combination of conserved and variable regions of the 16S rRNA gene (Kim *et al.*, 2011) can reveal a microbial community taxonomic profile.

Various online 16S rRNA gene sequence repositories and analysis tools allow researchers to compare experimental community profiles to those already known. Resources include SILVA, RDP, GreenGenes, NCBI, and NIH HMP (http://www.hmpdacc.org/). Analytical pipelines such as mothur (Schloss *et al.*, 2009) and QIIME (Caporaso *et al.*, 2010; Kuczynski *et al.*, 2011) provide a range of analysis options including diversity estimation and ordination methods. The eukaryotic domain of life is studied using a similar approach based on the same or different marker genes. For example, the 18S rRNA gene and the internal transcribed spacer (ITS) regions of the rRNA gene cluster have been used to study eukaryotic diversity (e.g. fungal, protozoan) in multiple environments (Penton *et al.*, 2013; Findley *et al.*, 2013).

Potential caveats of the marker-gene approach include problems deriving from the existence of dissimilar copies of 16S rRNA genes in some bacteria, which could artificially inflate the diversity statistics. Furthermore, this approach relies heavily on performance and robustness of the PCR primers targeting a region of interest. Finally, as the current next sequencing technologies cannot yet create sequences spanning the entire length of the 16S rRNA gene, the diversity estimation relies on an alignment of fragments of the 16S rRNA gene. Kim *et al.* (2011) report that different fragments of the 16S rRNA gene might be more representative of the diversity of archaea than bacteria. Coincidentally, since various research groups use different regions of the 16S rRNA gene in their diversity estimations, these reports become difficult to reconcile in the subsequent meta-analytic attempts.

Metagenomics – high-throughput shotgun (HTS sequencing) of microbial communities

While the marker gene-specific approach allows for the investigation of microbial diversity in an environment via the prism of a common marker gene-set, HTS metagenomic sequencing (also referred to as whole-genome shotgun or WGS) sequencing of bulk microbial community genomic DNA without the need for prior cultivation can produce an unbiased representation of the complete set of microbial genes that exist. The advantage of WGS is its inherent propensity to capture DNA from a broad spectrum of organisms. This includes prokaryotes, DNA-based viruses (including bacteriophages), as well as microeukaryotes such as fungi and protozoa. RNA viruses are also possible targets of metagenomic study. However, different preparations are required to preserve and obtain RNA viral samples for sequencing, such as use of random priming and amplification coupled to sequence-independent single-primer amplification (SISPA) (Djikeng *et al.*, 2008).

Since the completion of the first bacterial genome in 1995 (Fleischmann *et al.*, 1995) thousands of microbial genomes have been sequenced and methods for their assembly, prediction of genes and annotation has grown considerably (Petty, 2010). Since a single genome is examined in these investigations, it is possible with bioinformatic tools to construct a robust picture of the genome organization including genes, operons and regulatory control elements. In contrast, metagenomic sequence presents a fundamentally different analytical problem as by definition multiple organisms, present at different frequencies, are represented in a sequenced sample. As a result, even with deep sequencing, in most instances the metagenomic sequence recovered from a sample will be fragmentary and incomplete in nature with unequal distributions of genome coverage. Owing to these circumstances and the fact that many of the genomes present may not be represented with an annotated genome in the database, it is not always possible to identify the source or function of all sequence recovered. A further complication of metagenomic sequencing conducted from specimens derived from a biological host, such as in the case of microbiome studies, is the presence of host nucleic acids in the microbial community nucleic acid preparations and resulting sequence datasets. Although strides are being made in laboratory protocols which can reduce the amount of host in nucleic acid preparations, the amount of 'host contamination' still needs careful consideration when determining the depth of metagenomic sequence obtained per sample. Further, these 'host contaminant' reads require identification and removal by computational tools prior to subsequent metagenomic analysis. As a result of these challenges, the development and application of computational tools to facilitate metagenomic analysis and interpretation is continuing to evolve at a rapid pace.

There is no one correct methodology for analysis of metagenomic data. Once metagenomic sequence data are obtained, the steps taken will depend on the study design and questions to be answered. However, there are several analytical steps that are frequently taken to facilitate metagenomic investigations and these steps will be briefly introduced here. Many of these analytical techniques focus on the reduction of the metagenomic sequence dataset size, thus potentially improving the signal-to-noise ratio. Some of these techniques include sequence-quality threshold filtering and the collapse of redundant sequences at various levels of identity (Wooley *et al.*, 2010). Once the read set has been filtered, options exist to analyse individual reads and/or to apply assembly methods to place overlapping short sequences into longer 'contigs', or to apply a combination of both approaches (Human Microbiome Consortium, 2012a,b). Before gene-centric comparisons are carried out, careful attention needs to be given to potential biasing factors, including sample community composition, as average genome size affects gene frequencies (Beszteri *et al.*, 2010), and reference sequence length bias [or reads-count bias (Human Microbiome Consortium, 2012b)]. For analysis of individual reads, options include the use of the HMP Unified Metabolic Analysis Network (HUMAnN) software (Abubucker *et al.*, 2012) for reconstructing the metabolic potential of metagenomic samples, which proportionally weights reference gene families by the number of hits while taking alignment quality and reference sequence length into account.

A variety of assembly approaches can be used in the context of metagenomic analysis. Assembly has the advantage of examining longer contiguous sequences, thus increasing the likelihood of identifying regulatory control elements and full-length genes. DNA sequences can be *in silico* translated into amino acid sequences using gene calling approaches (Wooley *et al.*, 2010). Traditional assemblers written for whole genome assembly are often inadequate for metagenomic assembly (Nagarajan and Pop, 2010) and recently a variety of algorithms have been written in an attempt to improve this process [for example, MetaVelvet (Namiki *et al.*, 2012), SPAdes (Bankevich *et al.*, 2012), and Meta-IDBA (Peng *et al.*, 2011)]. However, as with many analysis steps associated with metagenomic analysis, this remains an active area of research.

An important goal of metagenomic analysis is to call open reading frames (ORFs), which can be *in silico* translated into amino acid sequences using 'gene caller' approaches (Noguchi *et al.*, 2008; Rho *et al.*, 2010). The identification of ORFs provides information on protein coding sequences and is the basic source for functional annotation. Further, the identification of ORFs serves to further diminish the total number of sequences to be analysed. Annotation of the experimentally derived sequences or ORFs ultimately relies on comparisons against curated repositories of known sequences [for example, UniRef (Suzek *et al.*, 2007), the National Center for Biotechnology Information Reference Sequence (RefSeq) database (Pruitt *et al.*, 2012), M5nr (Wilke *et al.*, 2012), and Kyoto Encyclopedia of Genes and Genomes (KEGG), (http://www.genome.jp/kegg/pathway.html)] for functional and taxonomic annotation. Typically two overarching approaches to annotation are pursued. The first is based on sequence homology in which a sequence alignment tool (such as BLAST) is used to identify best matches against such a database. As a single-input sequence can be a degenerate match for multiple references in the database, subsequent statistical analysis is required before a final taxonomy or functional assignment is derived. In the second approach, probabilistic models of biological sequence representing protein families can be constructed using hidden Markov models (HMMs) that provide accurate annotation and are amenable to high-throughput searches (Yoon, 2009).

Alternatively, and in lieu of gene calling, BLASTX can be used directly. As a single-input sequence can be a degenerate match for multiple references in the database, advanced gene family and organism 'counting' approaches take multiple hits into account. For example, MEGAN implements a least common ancestor approach that builds a taxonomic consensus based on multiple reference alignments (Huson *et al.*, 2007). Advanced phylogenetic inference methods, such as the automated pipeline for phylogenomic analysis (AMPHORA), can quickly identify dozens of universal single-copy marker genes to assign sequences to phylotypes and build genome trees (Wu and Eisen, 2008). Resulting gene family and organismal/phylotype frequencies can then be compared to identify differentially abundant features between distinct populations of samples. Additional methods for taxonomic binning of metagenomic reads and assemblies can also be conducted using high-throughput phylogeny and machine-learning based approaches such as the Automated Phylogeny Inference System (APIS) (Zeigler Allen *et al.*, 2012), which is a taxonomic classification system that uses a reference database to assign taxonomy to predicted proteins in a metagenomic dataset by constructing phylogenies using query metagenomic sequences and their neighbours in the database. Machine-learning approaches such as MGTAXA (Williamson *et al.*, 2012) build models using training data that can be used to assign taxonomy to metagenomic sequences using their k-mer signature.

Applications of metagenomics to the study of human health and disease

Since the development of microbiology as an applied science in the late 19th century including the pioneering work of Pasteur and Koch, it has been recognized that there are bacteria associated with the human body, which may play roles in health and disease. Throughout the 20th century, advancements in microbiology, genetics and medicine continued to strengthen our understanding of microbial associations with humans (Savage, 2001). While microorganisms are estimated to make up only 1–2% of the mass of the body of a healthy human, they are nonetheless estimated to outnumber human cells by 10 to 1. The term 'microbiome' was introduced to refer to the collection of microorganisms and their genomes associated with the human host. Humans (and all mammals) are complex assemblages of mammalian cells that are further organized into tissues and functional organs, constituting the basic structural units of the human body. In addition, bacterial cell populations are organized into communities that are perpetually associated with the human body throughout life. These communities are thought to enable human physiology by virtue of their participation in a multitude of interactions with their human hosts, including processes related to development, nutrition, immunity and resistance to pathogens. As such, the study of human biology cannot focus solely on human cells.

As with the investigation of any environment, understanding the inherent microbial community diversity is a critical first component for studying the human microbiome in order to elucidate the distribution and assembly patterns of microbial communities across body habitats and individuals, and to facilitate microbial ecological and biological studies. Soon after the use of 16S rRNA gene surveys to study microbial diversity in 'outside' environments, these methods were also applied to the human body (Kroes *et al.*, 1999; Eckburg *et al.*, 2005). With the advent of next-generation sequencing technologies the number of studies on the microbial diversity of the human body has continued to increase.

The earliest large-scale human microbiome studies using the 16S rRNA gene as a biomarker investigated various habitats of the human body including the gut (Eckburg *et al.*, 2005; Bik *et al.*, 2006; Gill *et al.*, 2006; Turnbaugh *et al.*, 2009), oral cavity (Aas *et al.*, 2005; Nasidze *et al.*, 2009) and skin (Grice *et al.*, 2009). These studies opened a new lens through which to view human health and disease and created the potential to reconsider or generate new preventative medical strategies and therapeutics. Along with elucidating a wide diversity of bacteria, collectively, these studies have suggested several common themes. First, the human body plays host to a variety of habitats from which different microbial community structures (presence and abundance) can be recovered. Further, these studies of relatively small numbers of sample first began to elucidate important concepts such as the nature and extent of a core microbiome, or a shared microbiota between individuals at different body sites (Turnbaugh *et al.*, 2009), and interpersonal variation in terms of phylotypes recovered (Costello *et al.*, 2009; Nasidze *et al.*, 2009; Turnbaugh *et al.*, 2009). In general, these studies first suggested that relatively few microbes are shared between individuals in gut and skin habitats and, although still small in number, a somewhat larger contingent of taxa have been reported to be shared in the oral cavity (Zaura, 2012). The potential presence of multiple core microbiomes associated with groups of individuals has also been suggested from studies of microbial communities associated with the vagina (Gajer *et al.*, 2012) and in studies of gut communities from the work of the MetaHit project (Qin *et al.*, 2010). These contrasting results suggested that further investigation of the core microbiome within and between individuals is warranted.

Prior to the HMP, the largest human metagenomic study undertaken has been the MetaHIT project, which has focused on WGS or metagenomic sequencing of whole community DNA profiles obtained from faecal specimens collected from 124 European individuals. This project produced a total of 576.7 Gb of sequence and led to the identification of 3.3 million non-redundant microbial genes, about 150-fold more than the human genome equivalent (Qin *et al.*, 2010). Additional examinations of these data introduced the concept of 'enterotypes' or clusters of individuals whose membership is characterized by specific, highly abundant microbiota (Arumugam *et al.*, 2011).

The HMP (NIH HMP Working Group *et al.*, 2009) was established by the NIH Common Fund (http://commonfund.nih.gov/hmp/) to provide a public resource to facilitate human microbiome research. To undertake the goals of the HMP, two main components of investigation have been pursued. The first has been to obtain a set of reference genomes (viral, bacterial and eukaryotic), which provide a critical framework for subsequent metagenomic annotation and analysis (Human Microbiome Jumpstart Reference Strains Consortium *et al.*, 2010). The second has focused on creating a baseline of microbial community structure and function (Human Microbiome Consortium, 2012a,b). These baseline datasets consist of both 16S and metagenomic sequence datasets from 300 subjects defined as 'healthy' individuals using a carefully defined set of clinical inclusion and exclusion criteria. A total of 15 (male) or 18 (female) body habitats have been investigated in five regions of the body (airways, skin, oral cavity, gastrointestinal tract, and vagina). These datasets represent the most comprehensive 16S and WGS profiles from a healthy adult population examined to date. More importantly, these datasets serve as a framework to probe the nature and extent of the microbial communities living in and on the human body of adult donors in an effort to better understand their role in human health and disease, thus providing a critical baseline for future metagenomic studies of the human microbiome. Further, the large sample size of this study affords an important opportunity to better understand both the presence and distribution of abundant versus rare taxa and metabolic functions. Both data types are necessary in order to establish a framework from which ecological and other controlling factors of the human microbiome can be elucidated.

In order to pursue the goals of the HMP it was necessary to enlist a multitude of scientific expertise (collectively referred to as the HMP Consortium) to develop, evaluate and standardize a variety of protocols from clinical sampling of specimens and extraction of genomic DNA, to the creation of reference genome sequence, and sequencing, annotation and analytical tools and protocols to generate and analyse 16S and WGS profiles (Human Microbiome Consortium, 2012a). A Data Analysis and Coordination Centre (DACC) was created to serve as the central repository for all HMP WGS, 16S, and reference genome sequence information generated by the four sequencing centres (Baylor College of Medicine, Broad Institute of MIT and Harvard, JCVI, The Genome Institute at Washington University). The reader is further referred to the DACC website (http://www.hmpdacc.org/) for additional information on all protocols and datasets produced by the HMP Consortium and detailed publications that have been generated from the Consortium. Additional projects supported by the HMP are investigating the association of specific components and dynamics of the microbiome with a variety of disease conditions, developing tools and technology including isolating and sequencing uncultured organisms, and studying the ethical, legal, and social implications of human microbiome research (http://commonfund.nih.gov/hmp/fundedresearch.aspx).

Among the central questions addressed by the HMP has been a further in-depth examination of the core or shared taxonomic and metabolic components of the microbiome across body habitats and individuals (Human Microbiome Consortium, 2012 a,b; Huse *et al.*, 2012; Li *et al.*, 2012, 2013). The ability to robustly examine this question has been facilitated by the large number of individuals and body sites sampled as part of the study design. The HMP study has confirmed the presence of high inter-individual variation and the importance of a rarefraction of organisms in these communities that serve as important reservoirs of genetic diversity (Human Microbiome Consortium, 2012a,b; Li *et al.*, 2012, 2013). In contrast to the findings suggested in the MetaHit study (Arumugam *et al.*, 2011), examination of the large HMP cohort did not appreciably replicate the findings of discrete enterotypes in the gut microbiome but rather the presence of smooth abundance gradients of dominant organisms (Human Microbiome Consortium, 2012b).

The concept of a functional core based on genes and pathways was preferentially supported in comparison to a taxonomic core. Therefore, although microbial taxonomic profiles varied between individuals, basic metabolic functions important to microbial life associated with the human body were consistently identified at all body sites (Human Microbiome Consortium, 2012b). While basic metabolic function was preserved across body habitats, the nature and extent of specific genes and pathways into which they are associated with varied. For example, the metabolism of simple sugars was enriched in the oral cavity. In contrast, the gut exhibited an overrepresentation of genes and pathways responsible for complex carbohydrate metabolism while the glycogen/peptidoglycan degradation was enriched in the vaginal region (Cantarel *et al.*, 2012). These findings suggest that the metabolic potential associated with body habitats of the healthy microbiome are shaped and balanced by a variety of environmental and genetic factors. For more information about the HMP, please consult the DACC website (http://www.hmpdacc.org/) and the HMP Collection (www.ploscollections.org/hmp) for additional in-depth analyses on HMP datasets.

Beyond the omes – systems biology views onto the host–microbiome interactions

From mammalian metagenomic studies it has been observed that individual hosts' microbiomes share a relatively constant and similar distribution of genes classified into functional categories (Cho and Blaser, 2012). The similarities exist despite profound variations of the taxonomic profiles across the hosts' inhabitants. To further understand this highly complex and functionally redundant system of organisms, a method is required to observe when and which specific functional modules become active.

An analysis analogous to a transcriptome study in humans, the metatranscriptomics of an environmental sample attempts to identify gene expression profiles. Such profiles can provide a snapshot of gene transcriptions at the level of the microbial community. Whereas metagenomics relies on robust and stable DNA molecules, metatranscriptomics utilizes RNA molecules extracted from microbial communities, which are turned into cDNAs via reverse transcription and then subjected to DNA sequencing (Wang *et al.*, 2009; Gilbert and Hughes, 2011). The process of sequencing cDNA to interrogate a transcriptome of either a single organism or community is often referred to as RNA-seq. Although microarrays (a venerable, previous generation alternative to RNA-seq) may still be a more accurate tool for expression profiling of known genes than the current adaptations of RNA-seq technologies, the microarray can only sample ecosystem genetic diversity using known gene sequences. Thus, an advantage of RNA-seq over microarrays is that it allows for discovery of yet-unclassified genes (Mäder *et al.*, 2011) and non-coding RNAs.

A metagenomic survey of a microbiome ultimately provides but a list of all genes – a map of all available parts that the microbiome can utilize to adapt and interact with the environment. Metatranscriptomics adds a dynamic layer to this map, highlighting which parts of the metagenome are active at a particular point in time and space (due to environmental exposure, for example). This becomes relevant when investigating host–microbiome interactions, as systems biology models can be constructed to include both host and microbiome components into more comprehensive interaction networks. Examination of the microbiome in the context of the host, and *vice versa* may lead to a better understanding of both systems, and ultimately to discovery of new biomarkers and therapies (Jumpertz *et al.*, 2011; Guo *et al.*, 2012). Systematic bioinformatics and correlation analysis strategies that integrate multiple streams of 'omics' data collected at different scales of time and levels of biological organization, leveraging comprehensive host metabolic and genotype information as well as environmental information from the human microbiome, will advance insights into disease pathogenesis (Fig. 10.1). Data integration and recursive feature selection will provide prioritized lists of biomarkers that can potentially deliver targets for diagnostic or predictive clinical tests (see below).

There are various ways to visualize molecular interactions/reactions, among them interaction graphs, pathway maps, and illustrative textbook figures. Each of these has its advantages and disadvantages. While interaction graphs are well suited to show all interactions of a certain molecule at a glance, they lack temporal and spatial information such as compartments. Pathway maps allow the visualization of a series of events but lack spatial information as well. Textbook figures provide spatial and temporal information, have the highest information content, and are thus best suited for summarizing biological concepts. An obvious limitation is that they cannot be automatically generated and their notation is inconsistent, often differing from author to author. Furthermore, this lack of standardization hampers data exchange and analysis. The challenge is to represent a series of molecular events with a rich set of unambiguous visual elements. Towards this end, one of the most advanced efforts in the systems biology field has been the development of the Systems Biology Graphical Notation and its integration with the Systems Biology Markup Language (SBML). The CellDesigner software (Funahashi *et al.*, 2007) and LibSBGN (van Iersel *et al.*, 2012) supports the automatic generation of graphical representations according to SBGN. Integrating metagenomic biomarker candidates within such zoom-able graphical maps encoding time, space and cellular state across individual species, will allow researchers to identify relevant context and develop a better understanding of underlying mechanisms.

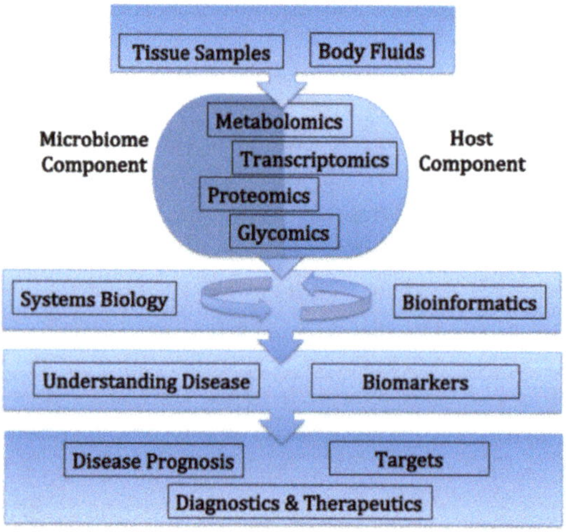

Figure 10.1 An overview of integrated systems biology approach interrogating data generated from multi-omic data streams that facilitate analyses featuring host and microbiome components.

The new generation of cloud-based informatics solutions for next-generation sequencing

Analysis of metagenomic sequencing data requires access to a range of technical capabilities including expertise in information technology (IT), bioinformatics, and software engineering, as well as significant computational and data storage capacity (Stein 2010). Sequencing data analysis approaches relying on large centralized facilities offering bioinformatics services to the community (Meyer *et al.*, 2008; Grigoriev *et al.*, 2012) are struggling to keep up with the volume and variety of data that is being generated. On the other hand, adoption of cloud computing technologies is providing an opportunity to change this paradigm by effectively democratizing the field, meaning that individual researchers and labs can now have access to the resources that were previously only available to large sequencing and informatics centres (Stein, 2010; Krampis *et al.*, 2012).

Along these lines, virtualization technologies allow entire computer servers including the operating system and all the necessary software packages for data analysis to be pre-installed, configured and ready-to-execute within a Virtual Machine (VM). A VM is an emulation of a computer server, with virtual processors, memory and storage capacity, in the form of a single binary file that can be executed on top of a hardware virtualization layer, independent of the underlying system architecture (Dudley and Butte, 2010; Dudley *et al.*, 2010). This approach can address one of the main hurdles to make open-source software with complex dependencies and installation procedures widely accessible to the research community. Consecutively, virtualization technologies led to the development of cloud computing services, where remote computer server farms can be rented on an hourly basis by researchers and used for scalable, on-demand computation. Cloud services offer high-performance computer hardware with a virtualization layer, upon which a user executes VM servers (Schatz *et al.*, 2010; Stein, 2010). Renting computational capacity from these services has the potential to eliminate many of the upfront capital expenses of building information technology infrastructure for microbial bioinformatics, and result in transformation of the analysis and data processing tasks into well-defined operational costs.

One platform bringing this technology into the bioinformatics field is Cloud BioLinux (Krampis *et al.*, 2012; http://www.cloudbiolinux.org) that supports building and distributing customized bioinformatics VMs on public clouds, and additionally sharing VM configurations between researchers that have their own private cloud installations. Cloud BioLinux provides the community with on-demand bioinformatics computing and a large assortment of pre-configured sequence analysis tools, packaged within a high-performance VM server that runs on private clouds and Amazon EC2. The project is targeted to researchers that do not have access to large-scale informatics infrastructure for their sequence data analysis needs, allowing users to rent computational capacity from cloud services and start the Cloud BioLinux VM by simply clicking through a web interface. The VM is open-source, can be downloaded and modified, while advanced users can run the tools on a private installation of the open-source Eucalyptus and OpenStack cloud platforms (http://open.eucalyptus.org, http://www.openstack.org). Cloud BioLinux has been featured on the official Amazon Web Services blog and Amazon Developer's newsletter, while a diverse community of researchers from both the US (e.g. Massachusetts General Hospital, Harvard School of Public Health, Emory University) and Europe (e.g. National Environmental Research Centre, King's College London, Denmark Technical University, Netherlands Wageninen University) has already been established around the project (https://groups.google.com/forum/#!forum/cloudbiolinux).

The Cloud BioLinux project to date has largely used the Amazon EC2 commercial cloud; however private and academic clouds based on the Eucalyptus and OpenStack cloud platforms have become commonplace computing provisions (Wilkening and Wilke, 2009). Other cloud computing resources include the government-funded Argonne National Lab Magellan Cloud (http://www.alcf.anl.gov/magellan) (Wilkening and Wilke, 2009), and the San Diego Supercomputer Center's cloud data storage system using Openstack (https://cloud.sdsc.edu/hp/index.php). We believe that this trend will continue forward, since in-house clouds can overcome the shortcoming of EC2 when large datasets are involved, related to bandwidth costs for transferring data to and from the cloud storage. For example, Eucalyptus and Openstack open-source cloud platforms can be readily deployed by an organization to run on internal servers.

Conclusion

As genome sequencing becomes cheaper, it is increasingly applied to sequence complex metagenomes. Large-scale 16S rRNA sequencing has become a routine practice for investigating prokaryote diversity in complex microbial communities. Similarly, WGS-based metagenomics has proven to be a powerful tool for surveying the complete genomic potential of communities. These technologies are increasing our understanding of the extent and role of microbial diversity in many ecologically important environments, with significant implications for understanding nutrient cycling, energy production and human health and disease. Potential challenges that require further development in the field of metagenomics studies include those associated with handling and storing large volumes of data, the standardization of assembly and analysis methods, and techniques for advanced graphical representations for data interpretation over time and space. To understand complex biological systems as an integrated whole, there is a need for more integrative studies which take into account the interconnected relationships between hosts, their associated microbes, and environmental factors.

Acknowledgements

This work is supported in part by Genome Sequencing Contract award funded by National Institute of Allergy and Infectious Diseases, National Institutes of Health, Department of Health and Human Services under contract number HHSN272200900007C and grant award funded by National Institute of Diabetes, Digestive and Kidney Diseases, National Institutes of Health, Department of Health and Human Services under award number DP3DK94343-01.

References

Aas, J.A., Paster, B.J., Stokes, L.N., Olsen, I., and Dewhirst, F.E. (2005). Defining the normal bacterial flora of the oral cavity. J. Clin. Microbiol. *43*, 5721–5732.

Abubucker, S., Segata, N., Goll, J., Schubert, A.M., Izard, J., Cantarel, B.L., Rodriguez-Mueller, B., Zucker, J., Thiagarajan, M., Henrissat, B., *et al.* (2012). Metabolic reconstruction for metagenomic data and its application to the human microbiome. PLoS Computl. Biol. *8*(6), e1002358.

Arumugam, M., Raes, J., Pelletier, E., Le Paslier, D., Yamada, T., Mende, D.R., Fernandes, G.R., Tap, J., Bruls, T., Batto, J.M., *et al.* (2011). Enterotypes of the human gut microbiome. Nature *473*, 174–180.

Bankevich, A., Nurk, S., Antipov, D., Gurevich, A.A., Dvorkin, M., Kulikov, A.S., Lesin, V.M., Nikolenko, S.I., Pham, S., Prjibelski, A.D., *et al.* (2012). SPAdes: a new genome assembly algorithm and its applications to single-cell sequencing. J. Comput. Biol. *19*, 455–477.

Berlemont, R., Pipers, D., Delsaute, M., Angiono, F., Feller, G., Galleni, M., and Power, P. (2011). Exploring the Antarctic soil metagenome as a source of novel cold-adapted enzymes and genetic mobile elements. Revista Argentina de Microbiologia *43*(2), 94–103.

Beszteri, B., Temperton, B., Frickenhaus, S., and Giovannoni, S.J. (2010). Average genome size: a potential source of bias in comparative metagenomics. ISME J. *4*, 1075–1077.

Bik, E.M., Eckburg, P.B., Gill, S.R., Nelson, K.E., Purdom, E.A., Francois, F., Perez-Perez, G., Blaser, M.J., and Relman, D.A. (2006). Molecular analysis of the bacterial microbiota in the human stomach. Proc. Natl. Acad. Sci. U.S.A. *103*, 732–737.

Brulc, J.M., Antonopoulos, D.A., Miller, M.E., Wilson, M.K., Yannarell, A.C., Dinsdale, E.A., Edwards, R.E., Frank, E.D., Emerson, J.B., Wacklin, P., *et al.* (2009). Gene-centric metagenomics of the fiber-adherent bovine rumen microbiome reveals forage specific glycoside hydrolases. Proc. Natl. Acad. Sci. U.S.A. *106*(6), 1948–1953.

Cantarel, B.L., Lombard, V., and Henrissat, B. (2012). Complex carbohydrate utilization by the healthy human microbiome. PloS one 7(6), e28742.

Caporaso, J.G., Kuczynski, J., Stombaugh, J., Bittinger, K., Bushman, F.D., Costello, E.K., Fierer, N., Peña, A.G., Goodrich, J.K., Gordon, J.I., *et al.* (2010). QIIME allows analysis of high-throughput community sequencing data. Nature Methods 7(5), 335–336.

Cho, I., and Blaser, M.J. (2012). The human microbiome: at the interface of health and disease. Nature Rev. Genet. *13*, 260–270.

Costello, E.K., Lauber, C.L., Hamady, M., Fierer, N., Gordon, J.I., and Knight, R. (2009). Bacterial community variation in human body habitats across space and time. Science *326*(5960), 1694–1697.

Daubin, V., Lerat, E., and Perrière, G. (2003). The source of laterally transferred genes in bacterial genomes. Genome Biol. *4*, R57.

Djikeng, A., Halpin, R., Kuzmickas, R., DePasse, J., Feldblyum, J., Sengamalay, N., Afonso, C., Zhang, X., Anderson, N.G., Ghedin, E., *et al.* (2008). Viral genome sequencing by random priming methods. BMC Genomics *9*, 5.

Dudley, J.T., and Butte, A.J. (2010). In silico research in the era of cloud computing. Nature Biotechnol. *28*, 1181–1185.

Dudley, J.T., Pouliot, Y., Chen, R., Morgan, A.A., and Butte, A.J. (2010). Translational bioinformatics in the cloud: an affordable alternative. Genome Med. *2*(8), 51.

Dupont, C.L., Rusch, D.B., Yooseph, S., Lombardo, M.J., Richter, R.A., Valas, R., Novotny, M., Yee-Greenbaum, J., Selengut, J.D., Haft, D.H., *et al.* (2012). Genomic insights to SAR86, an abundant and uncultivated marine bacterial lineage. ISME J. *6*(6), 1186–1199.

Durso, L.M., Harhay, G.P., Smith, T.P., Bono, J.L., Desantis, T.Z., Harhay, D.M., Andersen, G.L., Keen, J.E., Laegreid, W.W., and Clawson, M.L. (2010). Animal-to-animal variation in fecal microbial diversity among beef cattle. Appl. Environ. Microbiol. *76*(14), 4858–4862.

Eckburg, P.B., Bik, E.M., Bernstein, C.N., Purdom, E., Dethlefsen, L., Sargent, M., Gill, S.R., Nelson, K.E., and Relman, D.A. (2005). Diversity of the human intestinal microbial flora. Science *308*, 1635–1638.

Findley, K., Oh, J., Yang, J., Conlan, S., Deming, C., Meyer, J.A., Schoenfeld, D., Nomicos, E., Park, M.; NIH Intramural Sequencing Center Comparative Sequencing Program, *et al.* (2013). Topographic diversity of fungal and bacterial communities in human skin. Nature *498*, 367–370.

Fleischmann, R.D., Adams, M.D., White, O., Clayton, R.A., Kirkness, E.F., Kerlavage, A.R., Bult, C.J., Tomb, J.F., Dougherty, B.A., Merrick, J.M., *et al.* (1995). Whole-genome random sequencing and assembly of *Haemophilus influenzae* Rd. Science *269*(5223), 496–512.

Funahashi, A., Jouraku, A., Matsuoka, Y., and Kitano, H. (2007). Integration of CellDesigner and SABIO-RK. In silico Biol. 7 (Suppl. 2), S81–90.

Gajer, P., Brotman, R.M., Bai, G., Sakamoto, J., Schütte, U.M., Zhong, X., Koenig, S.S., Fu, L., Ma, Z.S., Zhou, X., *et al.* (2012). Temporal dynamics of the human vaginal microbiota. Sci. Translational Med. *4*, 132ra152.

Gilbert, J.A., and Hughes, M. (2011). Gene expression profiling: metatranscriptomics. Methods Mol. Biol. *733*, 195–205.

Gill, S.R., Pop, M., Deboy, R.T., Eckburg, P.B., Turnbaugh, P.J., Samuel, B.S., Gordon, J.I., Relman, D.A., Fraser-Liggett, C.M., and Nelson, K.E. (2006). Metagenomic analysis of the human distal gut microbiome. Science *312*, 1355–1359.

Grice, E.A., Kong, H.H., Conlan, S., Deming, C.B., Davis, J., Young, A.C.; NISC Comparative Sequencing Program, Bouffard, G.G., Blakesley, R.W., Murray, P.R., *et al.* (2009). Topographical and temporal diversity of the human skin microbiome. Science *324*, 1190–1192.

Grigoriev, I.V., Nordberg, H., Shabalov, I., Aerts, A., Cantor, M., Goodstein, D., Kuo, A., Minovitsky, S., Nikitin, R., Ohm, R.A., *et al.* (2012). The genome portal of the Department of Energy Joint Genome Institute. Nucleic Acids Res. *40*(Database issue), D26–32.

Guo, B., Harstall, C., Louie, T., Veldhuyzen van Zanten, S., and Dieleman, L.A. (2012). Systematic review: faecal transplantation for the treatment of Clostridium difficile-associated disease. Alimentary Pharmacol. Therapeut. *35*, 865–875.

Human Microbiome Consortium. (2012a). A framework for human microbiome research. Nature *486*, 215–221.

Human Microbiome Consortium. (2012b). Structure, function and diversity of the healthy human microbiome. Nature *486*, 207–214.

Human Microbiome Jumpstart Reference Strains Consortium, Nelson, K.E., Weinstock, G.M., Highlander, S.K., Worley, K.C., Creasy, H.H., Wortman, J.R., Rusch, D.B., Mitreva, M., Sodergren, E., *et al.* (2010). A catalog of reference genomes from the human microbiome. Science *328*(5981), 994–999.

Huse, S.M., Ye, Y., Xhou, Y., and Fodor, A.A. (2012). A core human microbiome as viewed through 16S rRNA sequence clusters. PLoS ONE 7(6), e34242.

Huson, D.H., Auch, A.F., Qi, J., and Schuster, S.C. (2007). MEGAN analysis of metagenomic data. Genome Res. *17*, 377–386.

van Iersel, M.P., Villéger, A.C., Czauderna, T., Boyd, S.E., Bergmann, F.T., Luna, A., Demir, E., Sorokin, A., Dogrusoz, U., Matsuoka, Y., *et al.* (2012). Software support for SBGN maps: SBGN-ML and LibSBGN. Bioinformatics *28*, 2016–2021.

Jumpertz, R., Le, D.S., Turnbaugh, P.J., Trinidad, C., Bogardus, C., Gordon, J.I., and Krakoff, J. (2011). Energy-balance studies reveal associations between gut microbes, caloric load, and nutrient absorption in humans. Am. J. Clin. Nutr. *94*, 58–65.

Kim, M., Morrison, M., and Yu, Z. (2011). Evaluation of different partial 16S rRNA gene sequence regions for phylogenetic analysis of microbiomes. J. Microbiol. Methods *84*, 81–87.

Krampis, K., Booth, T., Chapman, B., Tiwari, B., Bicak, M., Field, D., and Nelson, K.E. (2012). Cloud BioLinux: pre-configured and on-demand bioinformatics computing for the genomics community. BMC Bioinformat. *13*, 42.

Kroes, I., Lepp, P.W., and Relman, D.A. (1999). Bacterial diversity within the human subgingival crevice. Proc. Natl. Acad. Sci. U.S.A. *96*, 14547–14552.

Kuczynski, J., Stombaugh, J., Walters, W.A., González, A., Caporaso, J.G., and Knight, R. (2011). Using QIIME to analyze 16S rRNA gene sequences from microbial communities. Curr. Protoc. Bioinformatics, Chapter 10: Unit 10 17.

Li, K., Bihan, M., Yooseph, S., and Methé, B.A. (2012). Analyses of the microbial diversity across the human microbiome. PLoS ONE 7(6), e32118.

Li, K., Bihan, M., and Methé, B.A. (2013). Analyses of the stability and core taxonomic memberships of the human microbiome. PLoS ONE *8*(5), e63139.

Mäder, U., Nicolas, P., Richard, H., Bessières, P., and Aymerich, S. (2011). Comprehensive identification and quantification of microbial transcriptomes by genome-wide unbiased methods. Curr. Opin. Biotechnol. *22*(1), 32–41.

Meyer, F., Paarmann, D., D'Souza, M., Olson, R., Glass, E.M., Kubal, M., Paczian, T., Rodriguez, A., Stevens, R., Wilke, A., *et al.* (2008). The metagenomics RAST server – a public resource for the automatic phylogenetic and functional analysis of metagenomes. BMC Bioinformat. *9*, 386.

Nagarajan, N., and Pop, M. (2010). Sequencing and genome assembly using next-generation technologies. Methods Mol. Biol. *673*, 1–17.

Namiki, T., Hachiya, T., Tanaka, H., and Sakakibara, Y. (2012). MetaVelvet: an extension of Velvet assembler to de novo metagenome assembly from short sequence reads. Nucleic acids Res. *40*, e155.

Nasidze, I., Li, J., Quinque, D., Tang, K., and Stoneking, M. (2009). Global diversity in the human salivary microbiome. Genome Res. *19*(4), 636–643.

NIH HMP Working Group, Peterson, J., Garges, S., Giovanni, M., McInnes, P., Wang, L., Schloss, J.A., Bonazzi, V., McEwen, J.E., Wetterstrand, K.A., *et al.* (2009). The NIH Human Microbiome Project. Genome Res. *19*, 2317–2323.

Noguchi, H., Taniguchi, T., and Itoh, T. (2008). MetaGeneAnnotator: detecting species-specific patterns of ribosomal binding site for precise gene prediction in anonymous prokaryotic and phage genomes. DNA Res. *15*, 387–396.

Peng, Y., Leung, H.C., Yiu, S.M., and Chin, F.Y. (2011). Meta-IDBA: a de Novo assembler for metagenomic data. Bioinformatics *27*, i94–i101.

Penton, C.R., Stlouis, D., Cole, J.R., Luo, Y., Wu, L., Schuur, E.A., Zhou, J., and Tiedje, J.M. (2013). Fungal diversity in permafrost and tallgrass prairie soils under experimental warming conditions. Appl. Environ. Microbiol. *79*, 7063–7072.

Petty, N.K. (2010). Genome annotation: man versus machine. Nature Rev. Microbiol. *8*, 762.

Pruitt, K., Tatusova, T., Brown, G.R., and Maglott, D.R. (2012). NCBI Reference Sequence (RefSeq): current status, new features and genome annotation policy. Nucleic Acids Res. *40*, 130–135.

Pope, P.B., Mackenzie, A.K., Gregor, I., Smith, W., Sundset, M.A., McHardy, A.C., Morrison, M., and Eijsink, V.G. (2012). Metagenomics of the Svalbard reindeer rumen microbiome reveals abundance of polysaccharide utilization loci. PloS one *7*(6), e38571.

Qin, J., Li, R., Raes, J., Arumugam, M., Burgdorf, K.S., Manichanh, C., Nielsen, T., Pons, N., Levenez, F., Yamada, T., *et al.* (2010). A human gut microbial gene catalogue established by metagenomic sequencing. Nature *464*, 59–65.

Reddy, B.V., Kallifidas, D., Kim, J.H., Charlop-Powers, Z., Feng, Z., and Brady, S.F. (2012). Natural product biosynthetic gene diversity in geographically distinct soil microbiomes. Appl. Environ. Microbiol. *78*(10), 3744–3752.

Rho, M., Tang, H., and Ye, Y. (2010). FragGeneScan: predicting genes in short and error-prone reads. Nucleic Acids Res. *38*, e191.

Rusch, D.B., Halpern, A.L., Sutton, G., Heidelberg, K.B., Williamson, S., Yooseph, S., Wu, D., Eisen, J.A., Hoffman, J.M., Remington, K., *et al.* (2007). The Sorcerer II Global Ocean Sampling expedition: northwest Atlantic through eastern tropical Pacific. PLoS Biol. *5*(3), e77.

Savage, D.C. (2001). Microbial biota of the human intestine: a tribute to some pioneering scientists. Curr. Iss. Intestinal Microbiol. *2*(, 1–15.

Schatz, M.C., Langmead, B., and Salzberg, S.L. (2010). Cloud computing and the DNA data race. Nature Biotechnol. *28*(7), 691–693.

Schloss, P.D., Westcott, S.L., Ryabin, T., Hall, J.R., Hartmann, M., Hollister, E.B., Lesniewski, R.A., Oakley, B.B., Parks, D.H., Robinson, C.J., *et al.* (2009). Introducing mothur: open-source, platform-independent, community-supported software for describing and comparing microbial communities. Appl. Environ. Microbiol. *75*, 7537–7541.

Stein, L.D. (2010). The case for cloud computing in genome informatics. Genome Biol. *11*, 207.

Sun, Y., Cai, Y., Huse, S.M., Knight, R., Farmerie, W.G., Wang, X., and Mai, V. (2012). A large-scale benchmark study of existing algorithms for taxonomy-independent microbial community analysis. Briefings in Bioinformat. *13*(1), 107–121.

Suzek, B.E., Huang, H., McGarvey, P., Mazumder, R., and Wu, C.H. (2007). UniRef: comprehensive and nonredundant UniProt reference clusters. Bioinformatics 23, 1282–1288.

Tringe, S.G., Zhang, T., Liu, X., Yu, Y., Lee, W.H., Yap, J., Yao, F., Suan, S.T., Ing, S.K., Haynes, M., *et al.* (2008). The airborne metagenome in an indoor urban environment. PloS one 3(4), e1862.

Turnbaugh, P.J., Hamady, M., Yatsunenko, T., Cantarel, B.L., Duncan, A., Ley, R.E., Sogin, M.L., Jones, W.J., Roe, B.A., Affourtit, J.P., *et al.* (2009). A core gut microbiome in obese and lean twins. Nature *457*, 480–484.

Venter, J.C., Remington, K., Heidelberg, J.F., Halpern, A.L., Rusch, D., Eisen, J.A., Wu, D., Paulsen, I., Nelson, K.E., Nelson, W., *et al.* (2004). Environmental genome shotgun sequencing of the Sargasso Sea. Science *304*(5667), 66–74.

Wang, Z., Gerstein, M., and Snyder, M. (2009). RNA-Seq: a revolutionary tool for transcriptomics. Nature Rev. Genet. *1*, 57–63.

Whon, T.W., Kim, M.S., Roh, S.W., Shin, N.R., Lee, H.W., and Bae, J.W. (2012). Metagenomic characterization of airborne viral DNA diversity in the near-surface atmosphere. J. Virol. *86*(15), 8221–8231.

Wilke, A., Harrison, T., Wilkening, J., Field, D., Glass, E.M., Kyrpides, N., Mavrommatis, K., and Meyer, F. (2012). The M5nr: a novel non-redundant database containing protein sequences and annotations from multiple sources and associated tools. BMC Bioinformatics *13*, 141.

Wilkening, J., and Wilke, A. (2009). Using clouds for metagenomics: a case study. Cluster Computing and Workshops. CLUSTER'09. IEEE International Conference on, IEEE.

Williamson, S.J., Rusch, D.B., Yooseph, S., Halpern, A.L., Heidelberg, K.B., Glass, J.I., Andrews-Pfannkoch, C., Fadrosh, D., Miller, C.S., Sutton, G., *et al.* (2008). The Sorcerer II Global Ocean Sampling Expedition: metagenomic characterization of viruses within aquatic microbial samples. PloS one *3*(1), e1456.

Williamson, S.J., Allen, L.Z., Lorenzi, H.A., Fadrosh, D.W., Brami, D., Thiagarajan, M., McCrow J.P., Tovchigrechko, A., Yooseph, S., and Venter, J.C. (2012). Metagenomic exploration of viruses throughout the Indian Ocean. PLoS One 7, e42047.

Wooley, J.C., Godzik, A., and Friedberg, I. (2010). A primer on metagenomics. PLoS Computat. Biol. *6*(2), e1000667.

Wu, D., Wu, M., Halpern, A., Rusch, D.B., Yooseph, S., Frazier, M., Venter, J.C., and Eisen, J.A. (2011). Stalking the fourth domain in metagenomic data: searching for, discovering, and interpreting novel, deep branches in marker gene phylogenetic trees. PloS one *6*(3), e18011.

Wu, M., and Eisen, J.A. (2008). A simple, fast, and accurate method of phylogenomic inference. Genome Biol. *9*, R151.

Yildirim, S., Yeoman, C.J., Sipos, M., Torralba, M., Wilson, B.A., Goldberg, T.L., Stumpf, R.M., Leigh, S.R., White, B.A., and Nelson, K.E. (2010). Characterization of the fecal microbiome from non-human wild primates reveals species specific microbial communities. PloS one *5*(11), e13963.

Yoon, B.J. (2009). Hidden Markov Models and their Applications in Biological Sequence Analysis. Current Genom. *10*, 402–415.

Yooseph, S., Sutton, G., Rusch, D.B., Halpern, A.L., Williamson, S.J., Remington, K., Eisen, J.A., Heidelberg, K.B., Manning, G., Li, W., *et al.* (2007).

The Sorcerer II Global Ocean Sampling expedition: expanding the universe of protein families. PLoS Biol. 5(3), e16.

Zaura, E. (2012). Next-generation sequencing approaches to understanding the oral microbiome. Advanc. Dental Res. *24*, 81–85.

Zeigler Allen, L., Allen, E.E., Badger, J.H., McCrow, J.P., Paulsen, I.T., Elbourne, L.D., Thiagarajan, M., Rusch, D.B., Nealson, K.H., Williamson, S.J., *et al.* (2012). Influence of nutrients and currents on the genomic composition of microbes across an upwelling mosaic. ISME J. *6*, 1403–1414.

Index

www.ingramcontent.com/pod-product-compliance
Lightning Source LLC
LaVergne TN
LVHW071341210926
848486LV00003B/203